ELECTRONICS
principles
and applications

Graham Giller

SIGMA PRESS – Wilmslow, United Kingdom

First published in 1991 by

Sigma Press, 1 South Oak Lane, Wilmslow, Cheshire SK9 6AR, England.

British Library Cataloguing in Publication Data

A CIP catalogue record for this book is available from the British Library.

ISBN: 1-85058-151-7

Typesetting and design by

Sigma Hi-Tech Services Ltd

Printed in Malta by
Interprint Ltd.

Distributed by

John Wiley & Sons Ltd., Baffins Lane, Chichester, West Sussex, England.

Acknowledgement of copyright names

Within this book, various proprietary trade names and names protected by copyright are mentioned for descriptive purposes. Full acknowledgment is hereby made of all such protection.

PREFACE

This book is a revision of the one that I prepared in the summer of 1988. Its is a thorough examination of the subject at a level suitable for a student of BTEC level 3 or 'A' level Electronics. This is also suitable for the first few years of a degree course in Physics.

So that readers do not get lost in mathematics, and thereby conclude that Electronics is a confusing subject, I have made my working more explicit than is the norm. However, I do use the mathematics required for the subject, so I refer you to the texts in the bibliography when you come across some working which you do not understand

Having said all this, it is my hope that you will find, as I have, that Electronics is an interesting and enjoyable area to work in.

Graham Giller

CONTENTS

1

BASIC ELECTRICITY

1.1 Direct Current

The aim of this section is to introduce the reader to a mathematical analysis of direct current, which is an essential building block to the understanding of electronics. The important pieces are Kirchoff's laws, and Thévenin's and Norton's theorems, but a brief description of potential difference (voltage), current and Ohm's law is included for completeness.

1.1.1 The Coulomb Interaction, Current and Charge

It is a well known phenomenon that materials such as glass, polythene, and wool, can be forced to acquire some property, called electrical charge, which means that they will attract/repel other charged bodies.

Work by Franklin (1706-90) showed that the amount of charge was always conserved and that the charging of two bodies could be thought of as the separation of two types of charge (labelled positive and negative for convenience) of equal magnitude. All bodies initially carry charge +Q and -Q such that the total effect (+Q + -Q) is zero. However, we can mechanically separate these charges (by rubbing a glass rod with a silk cloth, etc.) leaving two equally, but oppositely, charged bodies.

Coulomb proposed that the force between two charges Q_1 and Q_2, separated a distance r in vacuo, was given by:

$$F = k \frac{Q_2 \, Q_2}{r^2}$$

(in the S.I. system of units $k = 1/4\pi\varepsilon_0$ & $\varepsilon_0 \approx 8.85 \times 10^{-12}$ F m^{-1})

The amount of charge, Q, is measured in coulombs.

Volta (1745-1827) produced the charge separation (with its related phenomena - forces, shocks, sparks) using a chemical device known as a Voltaic pile or, more simply, as a cell. A large number of cells connected in series is called a battery.

1

Since the Voltaic pile produced a separation of charges it must work against the mutual repulsion of like charges. The total amount of work a pile has the capacity to expend, per unit charge, is called the potential difference (or e.m.f. = electromotive force or voltage) of the cell. It is measured in volts.

Formally a pile with an e.m.f. of 1 volt will transfer 1 joule of energy to each coulomb of charge. (1 joule is the energy delivered by an engine working at 1 watt in 1 second).

i.e. $$U = QV \tag{1.1}$$

(U = energy, Q = charge, V = e.m.f.)

If two charged bodies are connected via an electrical conductor then, after some time, the charge vanishes. During this discharge various phenomena are detected. The 'nature' of the conductor changes. It appears to posses some property (called electrical current) which heats the wire and creates a magnetic field.

A simple experiment can be used to show that the current is due to moving charges. Conventionally, a positive current is created when positive charge flows to a negatively charged body from an positively charged body, or when negative charge flows to a positively charged body from a negatively charged body.

The rate of flow of charge past some point = current = rate of discharge of a charged body.

$$I = dQ/dt \tag{1.2}$$

I = current in amperes, t = time. The notation dQ/dt means the *rate of change of Q with respect to t*. The power delivered by a source is defined as the rate of change of stored energy.

$$P = dU/dt$$

∴ using (1.1) & (1.2)

$$P = IV \tag{1.3}$$

1.1.2 Ohm's Law

Simple experiments reveal that the magnitude of the current, which can be determined from the magnetic effect using a moving coil ammeter, due to one particular battery is dependent on the nature of the conductor used. Also, the size of the current through one particular conductor depends on the e.m.f. of the cell connected across it. The dependence is called Ohm's Law.

$$V = RI \qquad (1.4)$$

R is a constant, called the resistance, which is a property of the material connected across the battery. Sometimes the relationship is written in terms of the conductance, G = 1/R, i.e. I = GV. Many circuit elements (components) do not obey Ohm's law; they are called non-Ohmic or non-linear components. Metal wires and carbon rods are ohmic; but diodes, transistors and valves are non-Ohmic.

A light bulb is non-ohmic because it has a metal filament which heats up, and the resistance of a metal is found to increase with temperature. The heating of the filament is due to the current passing through it. Another way of looking at equation (1.4) is that it requires 1 joule per coulomb of charge to push 1 A through a 1 Ω resistor. There is power lost in the resistor! Combining (1.3) with (1.4) gives:

$$P = I^2R \qquad (1.5)$$
and
$$P = V^2/R \qquad (1.6)$$

Alternatively we could write:

$$P = I^2/G \qquad (1.7)$$
$$P = V^2G \qquad (1.8)$$

These equations relate the power dissipated by a resistor to the voltage dropped across it, or the current flowing through it.

1.1.3 Kirchoff's First Law

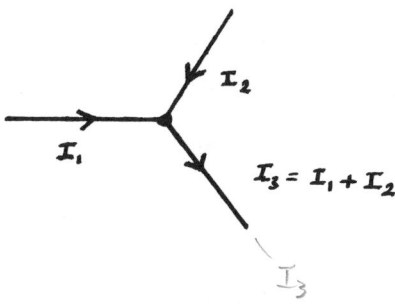

Figure 1.1 Kirchoff's 1st. Law

Kirchoff's first Law results from the continuity of charge. Electrical charge is a fundamental property of matter; we cannot create or destroy charge. Therefore, the total charge flowing into a circuit junction (a node) must equal the charge flowing out of it.

At a node: $$\sum I_i = 0 \qquad (1.9)$$

(The convention is that current flowing in is positive, while out is negative.)

1.1.4 Combination of Resistances

Figure 1.2 shows a circuit where two resistances, R_1 and R_2, have been connected in series.

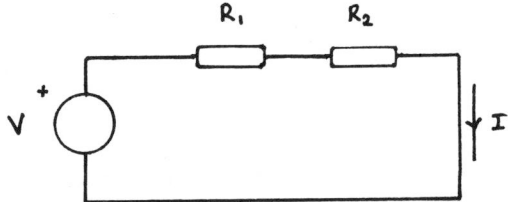

Figure 1.2 Series combination of resistances

A voltage, V, is applied across the circuit and this causes current, I, to flow through the circuit. It would be useful to find some equivalent resistance R' which we can use to replace the 'complicated' circuit.

$$V = IR'$$

by conservation of energy $$P = P_1 + P_2$$

$\Rightarrow$ $$I^2R' = I^2R_1 + I^2R_2$$

$\therefore$ $$R' = R_1 + R_2 \qquad (1.10)$$

Figure 1.3 shows a parallel combination of resistors. Here, the same potential is applied across both resistances.

$$V = I_1R_1 = I_2R_2$$

$\Rightarrow$ $$I_1 = VG_1 \text{ and } I_2 = VG_2$$

but using (1.9)

4

$$I' = I_1 + I_2 \text{ and also } I' = VG'$$

$\Rightarrow$

$$G' = G_1 + G_2$$

or

$$R' = \frac{R_1 R_2}{R_1 + R_2}$$

(1.11)

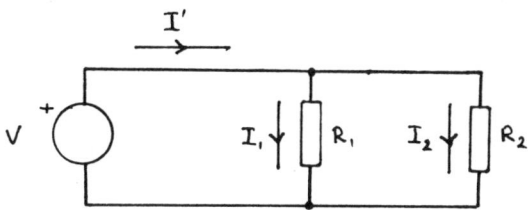

Figure 1.3 Parallel combination of resistances

1.1.5 The Voltmeter

The ammeter is a device which can measure current using the magnetic effect of a current, but there is nothing to measure voltages. However, if a resistance, R, is put across a potential, V, then the current flowing through it will be V/R. So, by measuring the current through a known load the voltage across the load can be determined. Suppose we want a meter with F.S.D. I amps to read V volts F.S.D. A resistance is put in series with the meter such that:

$$V = I(R + r)$$

(r = *internal resistance* of meter)

i.e.

$$R = V/I - r$$

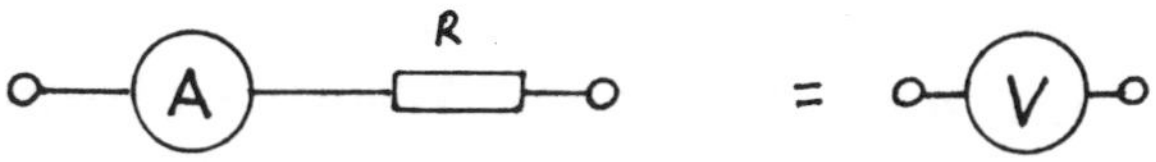

Figure 1.4 Moving coil voltmeter

This is how passive voltmeters are made. Now that we can measure voltages we can proceed to investigate the p.d.s in circuits.

1.1.6 Potential Dividers

A potential divider, or potentiometer or pot., is shown in figure 1.5. The arrow across the second resistor means that it is variable, i.e. its resistance can be altered using a sliding or rotary control.

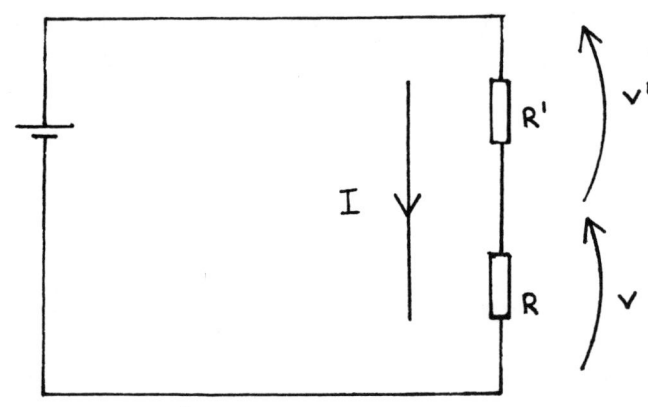

Figure 1.5 A potential divider

The voltage across the second resistor can be found using Kirchoff's and Ohm's law.

$$V = V_s \frac{R}{R + R'}$$ (1.12)

Also, note that the ratio of the potentials dropped by R and R' is equal to the ratio of their resistances.

$$V/V' = R/R'$$ (1.13)

1.1.7 Internal Resistance & Ideal Voltage Generators

Any real cell is found to behave as if it were an ideal e.m.f., which generates the same voltage across it whatever the current flowing through it, and a small series resistance. This is called the cell's internal resistance, or output resistance, and is usually written r.

The ideal voltage generator has a zero resistance.

If a moving coil voltmeter is used the e.m.f. of a cell can never be measured, because the measured voltage depends on the current flowing in the circuit, and this current depends on the resistance seen by the voltage generator. Figure 1.6 shows a practical attempt to measure the e.m.f. of a cell using a meter with total internal resistance R.

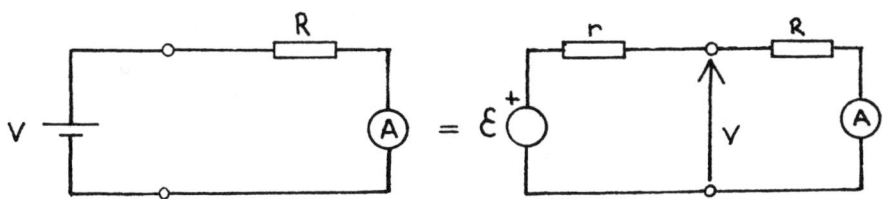

Figure 1.6 Effect of internal resistance on measured voltage

Obviously V_m is not equal to E.

$$V_m = ER/(R + r) \qquad (1.14)$$

From (1.14) it can be seen that the e.m.f. of the cell will be measured only when r is insignificant compared to R. Usually this condition is true, because r will be a few ohms whereas R will be much larger (kΩ) but, if it is not, care must be taken when measuring voltages.

1.1.8 Kirchoff's Second Law

Kirchoff's second law states that the total of the potentials measured across every component in any loop of a circuit is zero. This result derives from the conservation of energy.

around a loop $\qquad \sum V_i = 0 \qquad (1.15)$

Each e.m.f. generates a potential (taken to be positive), and each resistor drops a potential IR (taken to be negative).

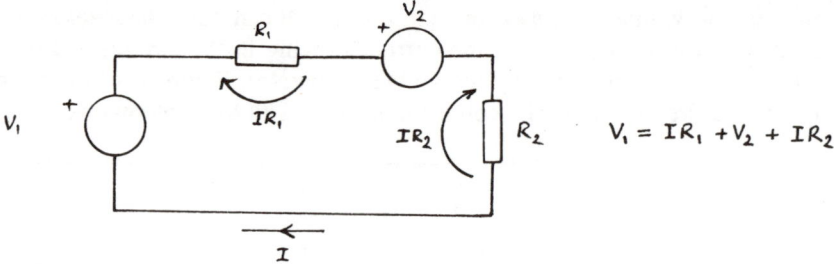

$V_1 = IR_1 + V_2 + IR_2$

Figure 1.7 Kirchoff's second Law

1.1.9 Thévenin's Theorem

AB is a complicated network involving e.m.f's and resistances.

We want to find the current which will flow through it when we connect an e.m.f. E and load R. To work this from first principles is a nightmare. Thévenin's theorem states that any two terminal networks of e.m.f.s and resistances can be represented as an e.m.f. E_0 in series with a resistance R_0. The e.m.f. E_0 is the potential which would be measured across AB when no current flows through load R, and the internal resistance R_0 is the resistance measured between AB when all e.m.f.s in the network are shorted.

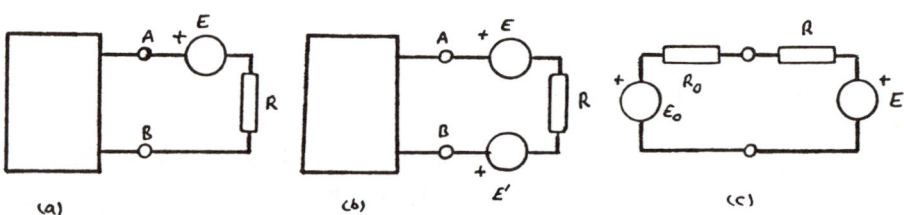

Figure 1.8 Proof of Thévinin's theorem

The theorem is easily proved. In the circuit external to the network we add an e.m.f. E' (b). We choose E' so that no current flows through the loop. This means that E' must

8

have the value:

$$E' = E - E_0$$

With all internal e.m.f.s short circuited, AB appears to be the resistance R_0 and, since Ohm's law is linear, the current flowing in (b) can be written:

$$I' = I - E' / (R + R_0)$$

By choosing I' to be zero we can use the value of E' above. Therefore the current in (a) is determined to be:

$$I = (E - E_0)/(R + R_0)$$

Which is what we would get if we used the equivalent (c).

1.1.10 Norton's Theorem

This is similar to Thévenin's theorem, but states that any network can be replaced with an ideal current source (develops no p.d., but causes a current to flow) in parallel with an internal resistance. To find I' short circuit the network at the terminals, and to find R' open circuit all current generators and calculate the resistance between the terminals.

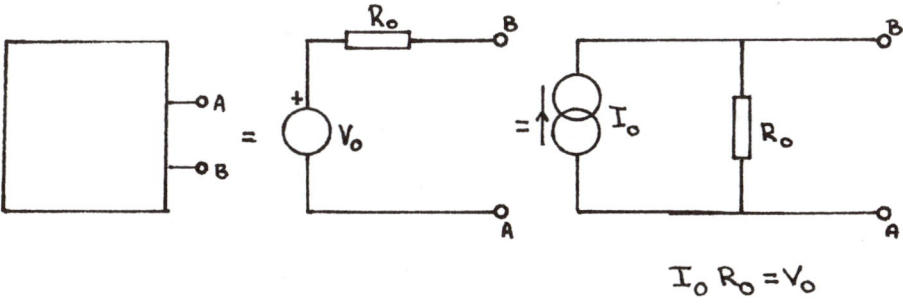

Figure 1.9 Thévinin and Norton equivalent circuits of a 'black box'

1.1.11 Transfer Theorems

There are three quantities that we may transfer from one part of a circuit to another: current, voltage, and power. From our discussion of internal resistance it should be obvious that we will get a maximum transfer of p.d. when the load approaches an open circuit, and that the current transfer is a maximum when the load approaches a short circuit.

For power transfer to a load, R, when the internal resistance, r, is fixed:

$$P = E^2R/(r + R)^2$$

$$\Rightarrow \quad\quad \left(\frac{\partial P}{\partial R}\right)_r = \frac{E^2}{(R + r)^2} - \frac{2RE^2}{(R + r)^3}$$

for a maximum/minimum this is zero $\Rightarrow$

$R \rightarrow \infty$ which obviously corresponds to minimum transfer

or $\quad\quad\quad\quad\quad$ R = r which is the maximum transfer if r is fixed.

For maximum power transfer to a load, when the load resistance is fixed, I^2R must be a maximum. With R fixed, this occurs when r = 0.

1.1.12 Representation of a Four Terminal Network

Using Thévenin's theorem any four terminal network can be represented by an input equivalent circuit and an output equivalent circuit.

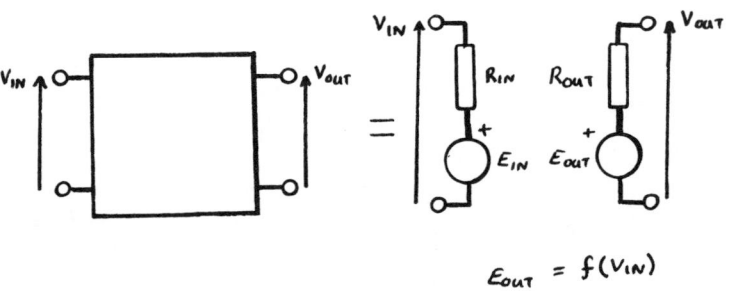

Figure 1.10 Thévinin equivalent circuit of a 4-terminal 'black box'

R_{in} is the input resistance and R_{out} is the output resistance. The output e.m.f. is generally some function of the input voltage, which is the p.d. measured across the input terminals, i.e. $E_{out} = f(V_{in})$. Such a representation can be used to calculate the actual current, voltage, and power transfer from circuit to circuit.

10

1.2 Alternating Current

This short section aims to introduce the concept of alternating current.

1.2.1 Electromagnetic Induction

Experiments done by Faraday on electromagnetic induction showed that when a charge is moved through a magnetic field (not parallel to it) it feels a force perpendicular to both the field and the direction of motion, and so moving a conductor with constant velocity through a field will cause the conductor to act as a d.c. current source.

However, such an arrangement is not practical. If a coil is rotated, such that at any time it has velocity v sin ωt perpendicular to the field (ω = angular velocity of coil), then the resulting current generated will also have a sinusoidal character. It is sometimes positive and sometimes negative, so it is described as alternating.

Since a rotating coil can be easily linked to some turbine system, such a.c. generators are far simpler to construct and use than d.c. generators (very large batteries). Mains electricity is a.c., and so we should study it. We should also study sinusoidal signals since a mathematical theorem allows any alternating current to be represented in terms of sinusoids, and the response of a circuit to sinusoids is simple to calculate.

1.2.2 Note on A.C. and Circuit Theorems

Over a small time interval the a.c. signal is approximately constant, and so we can apply the d.c. circuit theorems in that time interval. The a.c. signal is also approximately constant for the next time interval, so we can apply d.c. theorems in that interval too. So, the cycle can be divided up into small time intervals over which the d.c. theorems are true.

The d.c. theorems can be applied over the whole cycle when it is divided as above. If the length of the intervals is reduced any error which has been introduced, due to the approximations made above, should also reduce, until it vanishes as the length tends to zero, and we can conclude that the theorems are true over the entire cycle.

Note: The theorems cannot be said to be true when the wavelength of the a.c. signal is of the same order as the size of the circuit board. However, since this requires microwave frequencies (or higher), the error will be negligible for the circuits discussed here.

1.2.3 Measuring a Sinusoid

There are a number of important features which describe a sinusoid, outlined in the sketch.

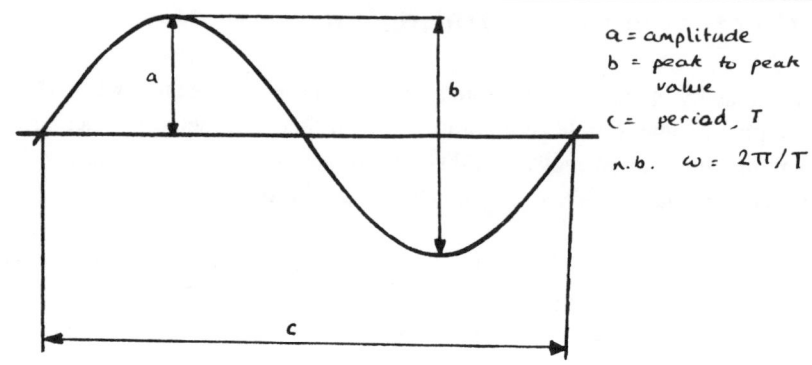

a = amplitude
b = peak to peak value
c = period, T

n.b. $\omega = 2\pi/T$

Figure 1.11 Sinusoid nomenclature

1.2.4 Root Mean Square Value of Signal

For an Ohmic component the current flowing through it due to the application of a sinusoidal signal is a sinusoid also.

at any time $$V = V_0 \sin \omega t$$

$\Rightarrow$ $$I = I_0 \sin \omega t$$ where $V_0 = I_0 R$

hence: $$P = I_0 V_0 \sin^2 \omega t$$

So, the instantaneous power delivered to the load is positive. Even though the mean p.d. across the load is zero, and the mean current through the load is zero, the load still dissipates power and so a.c. can be used to power appliances.

i.e. $$<V> = 0$$
$$<I> = 0$$

but $$<P> = \tfrac{1}{2} I_0 V_0$$

It is useful to find the d.c. voltage, V_{rms}, which will deliver the same power to the load as an a.c. signal of amplitude V_0.

i.e. $$V_{rms}^2 G = <V_0^2 \sin^2 \omega t> G$$

$$\Rightarrow \qquad V_{rms} = V_0/\sqrt{2} \qquad\qquad (1.16)$$

Note: The title 'root mean square' is used because we have taken the root of the mean of the square of the instantaneous amplitude.

1.3 Capacitors

This section studies the capacitor, a basic circuit element.

A capacitor is basically two metal plates which acquire a static charge when a cell is connected across them. In order to explain why they exist it is necessary to examine the structure of conductors, and to examine the structure of conductors it is necessary to examine the structure of atoms.

Every atom is electrically neutral since, if it were not, it would be impossible to produce a 'lump' of atoms of one kind. (The Coulomb repulsion between the atoms would force them apart!) Since it is possible to produce a block of pure lead, for example, the atoms must be neutral.

J J Thomson showed that all atoms contain a fundamental negatively charged particle, which is called an electron. Therefore, they must also contain a positively charged particle to make the atom neutral. Thomson proposed the 'plum-pudding' model of the atom in which negative electrons were embedded into a positive lump.

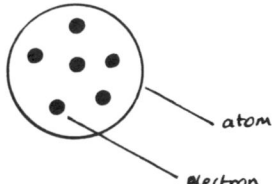

Figure 1.12 Model of the atom according to J J Thomson

Rutherford, as a consequence of work on alpha particles (small positive particles emitted from some radioactive substances), proposed an atomic model in which electrons orbit a very small and very dense positive nucleus. This model has a sound experimental basis, and there is no reason to doubt its validity.

Bohr proposed a more refined model in which the electrons were only allowed to exist in certain orbits about the nucleus, to explain the discrete nature of atomic spectra (light emitted from excited atoms contains only certain discrete frequencies, unlike that from a hot filament which contains every frequency up to a limit set by the temperature of the filament). This model, although considerably refined by the use of quantum mechanics, is accepted to be of the correct form.

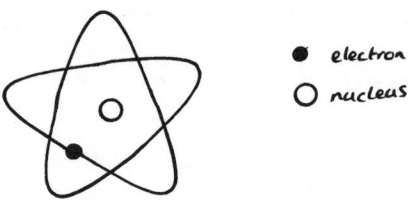

Figure 1.13 Model of the atom according to E. Rutherford

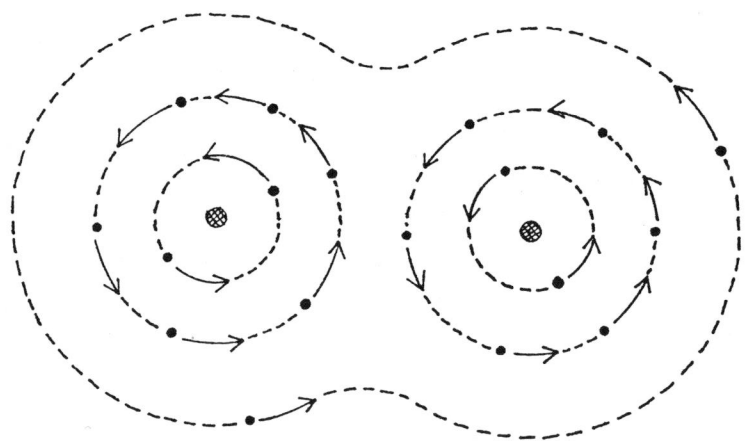

Figure 1.14 Model of diatomic molecule after atomic model of N Bohr

If we place two atoms close together the outermost orbits coincide, and so we might expect that the electrons could move from one atom to the other. This is indeed the case, and in a conductor very large numbers of atoms are close together so many electrons are free to move throughout the metal. If a potential is applied across the metal the electrons experience a force, due to the Coulomb interaction, and move towards the positive end of the metal. This means that the negative end of the metal acquires a slight positive charge (compared to the negative terminal of the cell) and electrons are drawn into the metal from the negative terminal of the cell. This leaves the cell with a positive charge, and so electrons from the metal flow into the cell to balance this. This is how a current flows in a conductor.

In a conductor the electrons separate as much as possible due to their mutual repulsion. At the negative terminal of a cell there is an excess of electrons (held there by the chemical reaction in the cell), and at the positive terminal there is a deficit of electrons. If conducting plates are connected to the terminals then the electrons will flow from the negative terminal onto the plate connected to it, and from the other plate onto the positive terminal, in order to increase the distance between the electrons to a maximum.

If you were to disconnect the plates now you would find that they have acquired equal, but opposite, electrical charges. This is how a capacitor charges up.

Commercial capacitors do not look like metal plates, but they are. If you were to examine one you would find two very thin metal foils, separated by an insulator (and rolled up).

Figure 1.15 Structure of a 'rolled foil' capacitor and symbol

---

1.3.1 Charge Stored on a Capacitor

By careful experimentation, the charges on plates of a capacitor can be found (e.g. by measuring the force between the plates). It is found that, for any particular capacitor, the charge is proportional to the potential applied across the plates.

$$Q = CV \tag{1.17}$$

The constant is dependent on the arrangement of plates and is called the *capacitance* of the capacitor.

1.3.2 Charge & Discharge of a Capacitor through a Resistive Load

Since we cannot produce an ideal voltage source let us investigate the charge and discharge through a load.

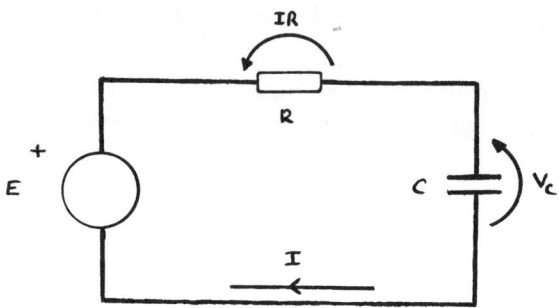

Figure 1.16 Charge up of a capacitor via series resistance

Suppose at some time during the charge-up the capacitor has charge Q

$$\Rightarrow \qquad V_C = Q/C$$

$$\therefore \qquad I = \frac{E - V_c}{R} \qquad (1.18)$$

Rewriting this in terms of charges gives a linear differential equation.

$$dQ/dt + Q/RC = E/R \qquad (1.19)$$

To solve this, make the equation homogeneous and then substitute a trial solution $Q = e^{kt}$

$$\Rightarrow \qquad (k + 1/RC)Q = 0$$

$$\Rightarrow \qquad \text{trivial solution } Q \equiv 0 \text{ or } k = -1/RC$$

The general solution of (1.19) is then:

$$Q(t) = Ae^{-t/RC} + B \qquad (1.20)$$

where A & B are arbitrary constants. The nature of B can be found by substitution into (1.19).

$$\Rightarrow \qquad \frac{Ae^{-t/RC}}{RC} - \frac{Ae^{-t/RC}}{RC} + B/RC = E/R$$

$$\Rightarrow \qquad B = EC$$

Now EC is the final charge that the capacitor will hold with e.m.f, E, across it. I will call this Q_0 for convenience.

$$\Rightarrow \qquad Q(t) = Ae^{-t/RC} + Q_0 \qquad (1.21)$$

All that remains is to extract A from the initial conditions.

For charging the capacitor $Q(0) = 0$:

$$\text{from } (1.21) \Rightarrow \qquad A = -Q_0$$

$$\therefore \qquad Q(t) = Q_0(1 - e^{-t/RC}) \qquad (1.22)$$

To solve for discharge we could start from scratch, or we could note that the discharge circuit is the same as the charge-up circuit with $E = 0$, which means $B = 0$.

$$\Rightarrow \qquad Q(t) = Ae^{-t/RC} \qquad (1.23)$$

The initial conditions are $Q(0) = Q_0$

$$\Rightarrow \qquad Q(t) = Q_0 e^{-t/RC} \qquad (1.24)$$

1.3.3 Capacitors and Complicated Networks

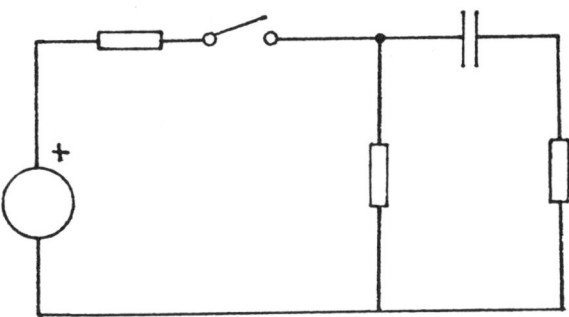

Figure 1.17 A complicated capacitor network

The simple circuits of §1.3.2 rarely occur in reality. What do you do if you are faced with a monster such as the network above, and have to find the charge on the capacitor as a function of time?

The answer is not to panic, just remember Thévenin's theorem: Redraw the network and then find the Thévenin equivalent e.m.f. E' and internal resistance R'. I have just solved this problem, and you should not waste time going through the working. Just write down the basic equation (1.19)

$$dQ/dt + Q/R'C = E'/R'$$

and the general solution equation (1.20)

$$Q(t) = Ae^{-t/R'C} + B$$

Then solve for B & A by substitution.

1.3.4 Capacitors and A.C.

Figure 1.18 illustrates a situation where I and V as a function of time must be found.

The solution will be of the form:

$$Q(t) = ae^{-t/RC} + b \cos \omega t + c \sin \omega t + d \qquad (1.25)$$

where $b \cos \omega t + c \sin \omega t$ is the particular integral of $V_0 \sin \omega t$.

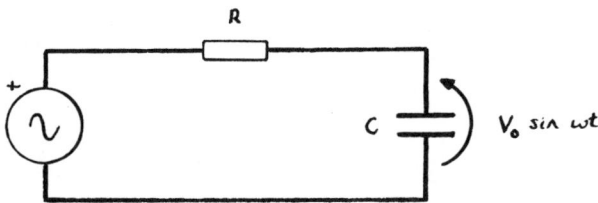

Figure 1.18 Response of a capacitor to a.c.

Apart from in a few isolated cases, we will not be interested in the *transient* part $ae^{-t/RC}$, nor the constant part d, but in the *steady state* part $b \cos \omega t + c \sin \omega t$. (Steady state means what is left as $t \rightarrow \infty$.)

If $V = V_0 \sin \omega t$ then $\qquad\qquad b = 0 \; \& \; c = V_0 C$

$\therefore \qquad\qquad\qquad\qquad V(t) = V_0 \sin \omega t \qquad\qquad\qquad (1.26)$

$\& \qquad\qquad\qquad\qquad I(t) = V_0 C \omega \cos \omega t \qquad\qquad\qquad (1.27)$

So the a.c. flows through the capacitor, with a change in current amplitude and a phase lag introduced between p.d. and current. These results will be discussed in a later section.

1.3.5 Combination of Capacitors

C' is the capacitor which can be used to replace C_1 and C_2 in the parallel circuit of figure 1.19.

The same p.d. is applied across each capacitor so:

$$Q_1 = C_1 V \text{ and } Q_2 = C_2 V \qquad\qquad (1.28)$$

Since the equivalent must take the same charge from the circuit:

$$Q = Q_1 + Q_2 = C' V \qquad\qquad (1.29)$$

$\Rightarrow \qquad\qquad\qquad\qquad C' = C_1 + C_2 \qquad\qquad\qquad (1.30)$

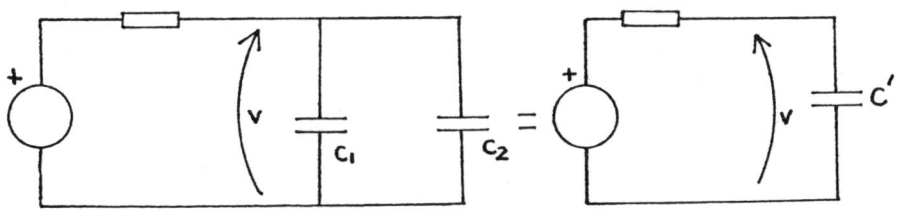

Figure 1.19 Parallel combination of capacitors

For the series circuit we must have:

$$V = V_1 + V_2 \qquad\qquad (1.31)$$

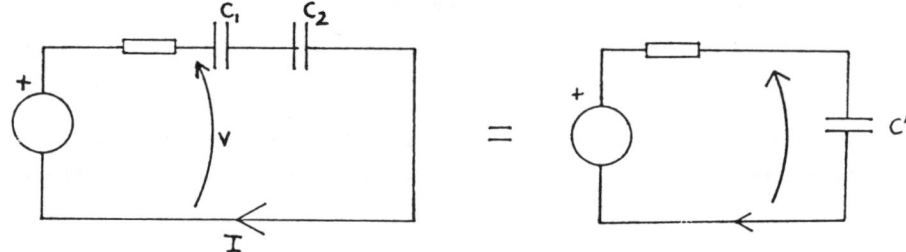

Figure 1.20 Series combination of capacitors

But, since the charging current is the same for each capacitor:

$$dQ_1/dt = dQ_2/dt \qquad (1.32)$$

integrating wrt. time $\Rightarrow$ $\qquad Q_1 = Q_2 + k$

Demanding an initial condition $Q_1(0) = 0 = Q_2(0)$ means that the constant vanishes.

The same current must charge the equivalent:

$\therefore \qquad\qquad\qquad dQ/dt = dQ_1/dt$

$\Rightarrow \qquad\qquad\qquad Q = Q_1 = Q_2$

and from (1.32) $\qquad\qquad 1/C' = 1/C_1 + 1/C_2 \qquad (1.33)$

1.3.6 Energy Stored in a Capacitor

For the circuit of figure 1.21, the energy required to transfer a small charge from one plate to another is:

$$dU = V \, dQ$$

Using (1.17) $\qquad\qquad C \, dU = Q \, dQ$

$\Rightarrow \qquad\qquad\qquad CU = \tfrac{1}{2}Q^2 + k$

The constant must be zero because an uncharged capacitor has no energy stored upon it.

Using (1.17) again $\Rightarrow$ $\qquad\qquad$ $U = \frac{1}{2}CV^2$ $\qquad\qquad\qquad$ (1.34)

Figure 1.21 Circuit used to calculate energy stored on a capacitor

1.4 Inductors

It is a lot harder to explain the operation of an inductor than a capacitor, even though the theory is simpler.

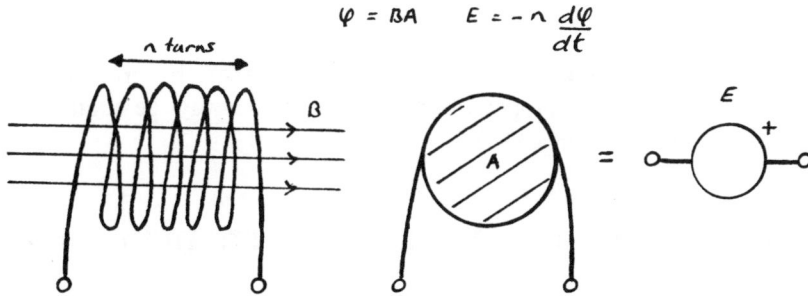

Figure 1.22 Illustration of Faraday's Law

Faraday's law of electromagnetic induction is:

$$E = -n \, d\varphi/dt$$

Which is the e.m.f. induced in a coil of n turns when the magnetic flux cutting it is changing at a rate $d\phi/dt$. The flux of a vector field (such as the magnetic field B) through a surface is hard to explain without vector calculus. However, for a uniform field, perpendicular to a flat surface of area A, we can use:

$$\phi = BA = \int_s \mathbf{B} \cdot d\mathbf{S}$$

as a definition of the flux of B (magnetic field strength) through the surface.

The magnetic field can be provided by the current flowing through the coil itself! From Ampere's law or the Biot-Savart law we can find the magnetic field strength due to a coil, which will generally be given by:

$$B = fI$$

where the form of f is determined by the nature (dimensions, no. turns, etc.) of the coil.

$\Rightarrow$ $\qquad\qquad\qquad\qquad E = - Afn \, dI/dt$

Changing the current in a coil generates an e.m.f. to oppose that change!

For an inductor, which is any wound component which has one coil, we define a quantity L, which is called the *inductance* of the component and is dependent on the nature of the coil, and so we get:

$$E = - L \, dI/dt \qquad\qquad (1.35)$$

In electronics, this is the equation that matters, not the specific form of L. It is a constant the value of which is a property of the coil and will always be quoted (just like capacitance). The unit of inductance is the Henry, H.

1.4.1 The Charge and Discharge of an Inductor

Although the inductor is an e.m.f. in circuit it is more useful to think of it as dropping a p.d. of -E in circuit.

Applying Kirchoff's second law to the circuit of figure 1.23:

$$E - IR - L \, dI/dt = 0$$

$\Rightarrow$ $\qquad\qquad\qquad\qquad dI/dt + IR/L = - E/L \qquad\qquad (1.36)$

This equation can be solved but, since the actual equation is identical in form to that for the capacitor, the solution can just be written down.

$$I(t) = Ae^{-tR/L} + B \qquad (1.37)$$

So there is an exponential solution for a *charge-up* (growth of current).

$$I(t) = I_0(1 - e^{-tR/L})$$

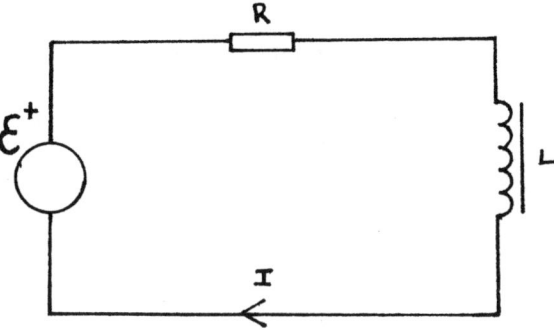

Figure 1.23 An LR series circuit

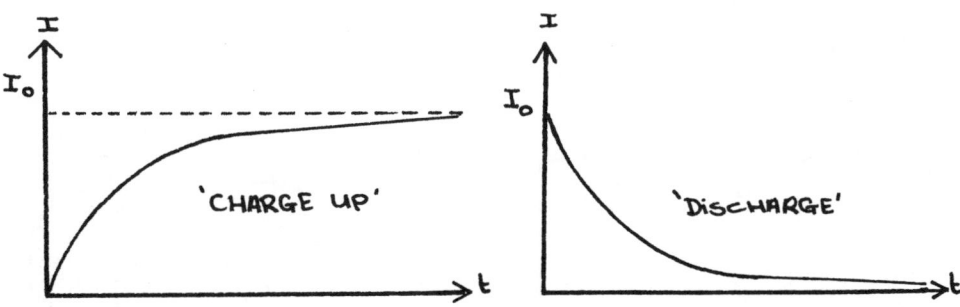

Figure 1.24 Growth and Decay of current in an LR circuit

I_0 is the steady state current, i.e. what you get when you replace the inductor by a short.

There is also a solution for *discharge* (decay of current).

$$I(t) = I_0 e^{-tR/L}$$

1.4.2 Inductors and A.C.

Here we have a similar situation to that in §1.3.4. We wish to find the current through the inductor when the LR circuit is driven by a sinusoidal e.m.f.

i.e.
$$dI/dt + IR/L = \frac{V_0 \sin \omega t}{L}$$

The steady state solution is:

$$I = A \sin \omega t + B \cos \omega t$$

$$V = A\omega \cos \omega t - B\omega \sin \omega t$$

If the p.d. across the inductor is $(V_0 \sin \omega t)/L$ then:

$$A = 0 \text{ and } B = - V_0/\omega L$$

As with the capacitor, there is a phase difference between current and voltage introduced and the magnitude of the current is changed.

1.4.3 Combinations of Inductors

For the series combination:

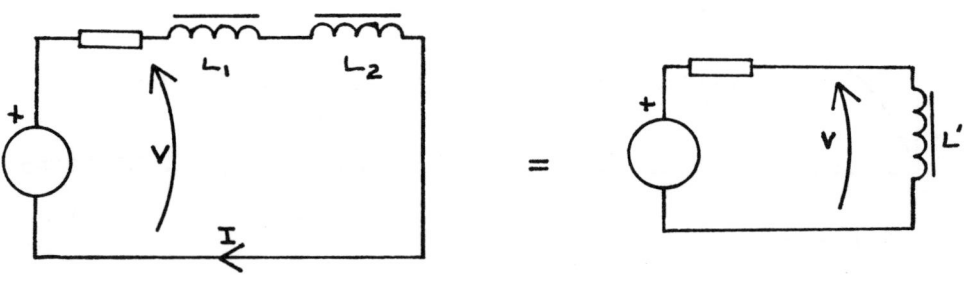

Figure 1.25 Series combination of inductances

$$V = L_1 \, dI/dt + L_2 \, dI/dt$$

and
$$V = L' \, dI/dt$$

$$\Rightarrow \qquad\qquad L' = L_1 + L_2 \qquad\qquad (1.38)$$

For the parallel combination:

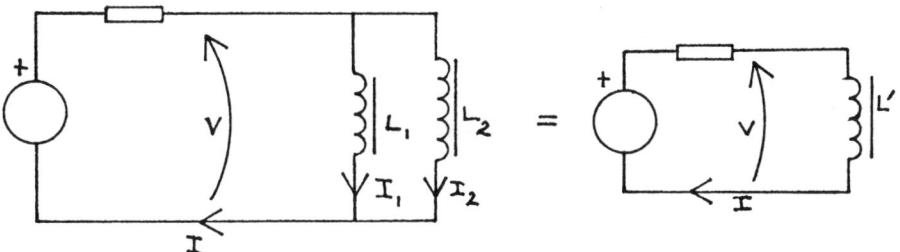

Figure 1.26 Parallel combination of inductances

$$V = L_1 \, dI_1/dt = L_2 \, dI_2/dt = L' \, dI/dt$$

where
$$I = I_1 + I_2$$

$$\Rightarrow \qquad\qquad dI/dt = dI_1/dt + dI_2/dt$$

$$\Rightarrow \qquad\qquad 1/L' = 1/L_1 + 1/L_2 \qquad\qquad (1.39)$$

1.4.4 Energy Stored in an Inductor

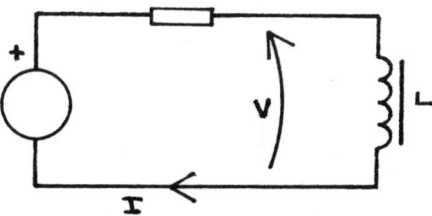

Figure 1.27 Circuit used for calculation of energy stored in an inductor

The power dissipated by the inductor is:

$$P = IV$$

$$\Rightarrow \qquad\qquad dU/dt = IL \, dI/dt$$

25

$$\therefore \qquad U = \tfrac{1}{2}LI^2 + k$$

The constant is zero, because there can be no energy stored in the inductor when there is no current flowing (the energy is stored in the magnetic field, which is present only when current flows).

$$\Rightarrow \qquad\qquad U = \tfrac{1}{2}LI^2 \qquad\qquad\qquad (1.40)$$

1.4.5 Driving Inductive Loads

The equation:
$$E = - L \, dI/dt$$

has serious implications if you are driving inductive loads such as motors, relays, or loudspeakers.

If you suddenly turn such a load off (and some solid state circuits can turn off in less than a nanosecond = 10^{-9} s) then a very large e.m.f. will be generated by the inductor. This is often sufficient to damage your semiconductor driver, or to cause sparks across metal contacts (which are a fire hazard and can also damage the contacts). Although, typically, the energy stored in the inductor will be slight if it is all released in a millisecond (for example), the power dissipated will be quite large.

1.5 Transformers

If we form an inductor, by winding a coil, and then drive alternating current through the coil, an alternating magnetic field is created. A second coil can be wound such that some of the flux created by the first coil cuts the second coil. According to Faraday's law, there will now be e.m.f. induced in the second coil, i.e.

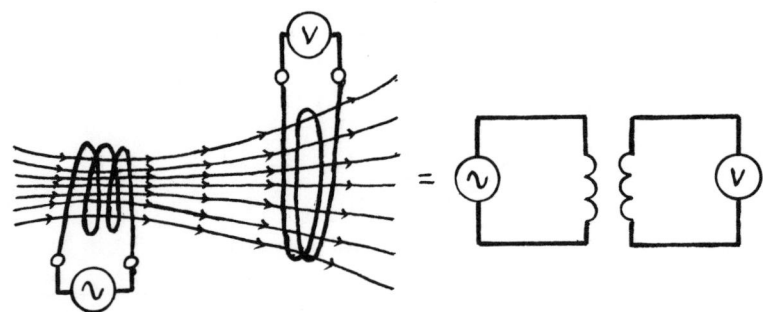

Figure 1.28 Linkage of magnetic field between circuits

This is a transformer.

1.5.1 Basic Transformer Equations (ratio rules)

The *primary* coil, with N_p turns, is driven from an e.m.f. $V_p \sin \omega t$. From Faraday's law:

$$N_p \, d\varphi_p/dt = V_p \sin \omega t \qquad (1.41)$$

Suppose some fraction of the flux in the primary coil, φ_p, is linked to the *secondary* coil, such that:

$$\varphi_s = k\varphi_p \qquad (1.42)$$

Then
$$d\varphi_s/dt = k \, d\varphi_p/dt$$

$$\Rightarrow \qquad \frac{d\varphi_s}{dt} = k \frac{V_p \sin \omega t}{N_p} \qquad (1.43)$$

Faraday's law must also apply to the secondary coil.

$$\therefore \qquad N_s \, d\varphi_s/dt = - E \qquad (1.44)$$

From (1.43) and (1.44) it is deduced that E, the e.m.f. generated in the secondary coil, is also a sinusoid of frequency $\omega / 2\pi$.

$$\Rightarrow \qquad E = - k \frac{N_s}{N_p} V_p \sin \omega t \qquad (1.45)$$

It can be written:
$$E = - V_s \sin \omega t \qquad (1.46)$$

Transformers are wound on an iron core. This is because the iron increases the magnetic field created by a factor of as much as 5000 or more, compared with an air cored device. You can think of the iron as offering a *low resistance* path to the magnetic field, and almost all of the flux is found to exist within the iron. Thus the *coupling* of the coils is usually very good. For ideal coupling $k = 1$.

Combining (1.45) and (1.46) gives the turns ratio rule for voltages

$$V_s/V_p = N_s/N_p \qquad (1.47)$$

Thus if $N_s = 2N_p$, for example, then $V_s = 2V_p$. Such a device is called a *step up transformer*, and if $N_s < N_p$ it is called a *step down transformer*. There must be conservation of energy so that power made available at the secondary can never exceed that consumed in the primary. With an ideal transformer:

$$P_p = P_s$$

$$\Rightarrow \qquad I_p V_p = I_s V_s$$

Which, by substitution into (1.47), gives the turns ratio rule for currents:

$$I_s/I_p = N_p/N_s \tag{1.48}$$

From these equations it is immediately obvious how a transformer can be used. Electricity is delivered to our houses at 240 V a.c. (rms), but this is of little use in electronic circuits where we usually need a maximum of 30 V d.c. A transformer can be used to change from 240 V a.c. to 30 V a.c. (To *rectify* the signal requires a diode circuit covered later.)

1.5.2 The Mutual Inductance of a Transformer

The equations (1.47) and (1.48) are useful for calculating the turns required to, for example, *step down* 240 V a.c. to 5 V a.c. in a power supply. However, an expression such as (1.34) is required for full network analysis with transformers.

The mutual inductance of the transformer is defined using the equation:

$$\varphi_s = MI_p \tag{1.49}$$

giving:

$$E_s = - M \, dI_p/dt \tag{1.50}$$

or

$$V_s = M \, dI_p/dt \tag{1.51}$$

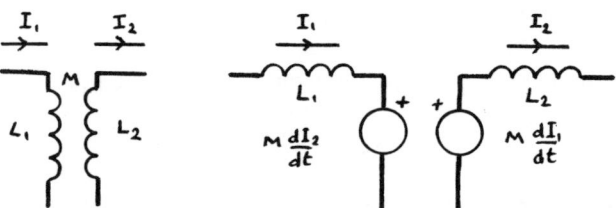

Figure 1.29 Treatment of Mutual Inductance

and then proceed just as for self inductance, L. (M is also given in Henrys). These equations also apply for the dependence of primary e.m.f.'s on secondary currents, and the form can be obtained by swapping the labels 's' and 'p'.

i.e. $$V_s = M \, dI_p/dt \quad \text{and} \quad V_p = M \, dI_s/dt$$

The equations take the form written here when the primary and secondary currents are measured in the same direction (such as left to right). If they are measured in opposite directions (left to right in the primary and right to left in the secondary, etc.) then 'M' should be replaced by '-M'.

As an example, I shall treat the coupled circuit of figure 1.30.

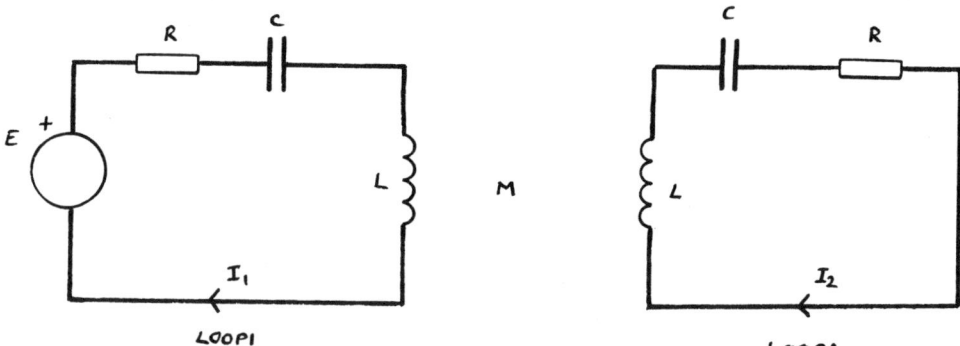

Figure 1.30 An 'LCR-LCR' coupled circuit

for loop 1 (primary):

$$L \, d^2Q_1/dt^2 + R \, dQ_1/dt + Q_1/C = E + M \, d^2Q_2/dt^2$$

for loop 2 (secondary):

$$L \, d^2Q_2/dt^2 + R \, dQ_2/dt + Q_2/C = M \, d^2Q_1/dt^2$$

These must be solved simultaneously.

As the example shows, both the self and mutual inductances generate circuit e.m.f.'s. A transformer formed from two inductors, L_1 and L_2, has the mutual inductance, M, given by equation (1.52).

$$M = k\sqrt{(L_1 L_2)} \tag{1.52}$$

k is called the *coefficient of coupling*, and for a *tightly* coupled coil (i.e. a practical, if not theoretical, ideal) is ≈ 0.5.

1.6 The Complex Impedance

This is an important section. It deals with the idea of the complex impedance.

1.6.1 Comparing Sine Waves

When we are comparing sine waves of the same frequencies there are only two numbers we have to know. One is the relative amplitudes and the other is the relative phases.

for $y = A \sin (\omega t + \varphi)$; A = amplitude, φ = phase angle

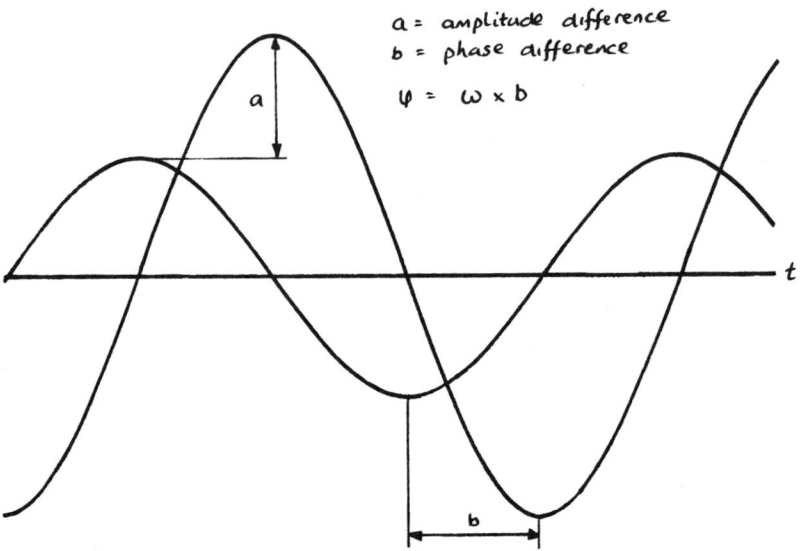

a = amplitude difference
b = phase difference
$\varphi = \omega \times b$

Figure 1.31 Comparing two sinusoids with common frequency

1.6.2 The Complex Impedance

If you refer back to §1.2.4, §1.3.4 & §1.4.2 you will find that, if the current wave through a component is defined to have zero phase,

i.e. $\qquad\qquad\qquad\qquad I = I_0 \sin \omega t \qquad\qquad\qquad\qquad (1.53)$

and the voltage drop across any component is calculated, the results are:

30

for a resistor $V = RI$:　　　　　　　　$V = RI_0 \sin \omega t$

for a capacitor $V = \int I/C \, dt$:　　　　$V = -(I_0 \cos \omega t)/\omega C$　　　　　　(1.54a,b,c)

for an inductor $V = L \, dI/dt$:　　　　$V = \omega L I_0 \cos \omega t$

Using the relationships: $- \cos a \equiv \sin (a - \tfrac{1}{2}\pi)$, $\cos a \equiv \sin (a + \tfrac{1}{2}\pi)$

for a resistor:　　　　　　　　　$V = RI_0 \sin (\omega t + 0)$

for a capacitor:　　　　　　　　$V = \{(I_0 \sin (\omega t - \tfrac{1}{2}\pi)\}/\omega C$　　　(1.55a,b,c)

for an inductor:　　　　　　　　$V = \omega L I_0 \sin (\omega t + \tfrac{1}{2}\pi)$

So, the action of a resistor is to drop a p.d. RI_0 in phase with the current. The action of a capacitor is to drop a p.d. $I_0/\omega C$ with relative phase $-\tfrac{1}{2}\pi$, and the action of an inductor is to drop a p.d. $\omega L I_0$ with relative phase $\tfrac{1}{2}\pi$.

The complex number system is used to express these results concisely.

A complex current is defined:　　$\mathbf{I} = I_0 e^{j\omega t}$　　　　　　　　(1.56a,b)
and voltage:　　　　　　　　　　$\mathbf{V} = V_0 e^{j(wt + \delta)}$

The actual current and voltage are the imaginary parts of the complex currents. A complex version of Ohm's law can be used to relate V to I for any of the components discussed in this chapter. That law is:

$$\mathbf{V} = \mathbf{ZI} \quad \text{or} \quad \mathbf{Z} = \mathbf{V}/\mathbf{I} = (V_0/I_0)e^{j\delta} \qquad (1.57)$$

which gives for a resistor:　　　　$\mathbf{Z}_R = R$

for a capacitor:　　　　　　　　$\mathbf{Z}_C = -j/\omega C$　　　　　　　(1.58a,b,c)

and for an inductor:　　　　　　$\mathbf{Z}_L = j\omega L$

If these complex representations are used, the phase changes will be automatically taken care of when the the modified form of Ohm's law is applied. Complex analogies to the network theorems can also be used. For a general component, which is made from a network of capacitors, inductors, and resistors, its complex impedance can be expressed as:

$$\mathbf{Z} = R + jX \qquad (1.59)$$

Where R is its *resistance* and X is its *reactance*. You will see that a resistor is a purely resistive component (hence its name), and capacitors and inductors are purely reactive components. The impedance is the modulus of the complex impedance, and the phase angle is the argument of the complex impedance.

i.e.
$$Z = |Z| = (R^2 + X^2)^{\frac{1}{2}} \qquad (1.60)$$

and
$$\delta = \arg Z = \tan^{-1} X/R$$

With these results in mind we now can find the voltage dropped across any component network thus:

1   Find the complex impedance of the network.

2   Assuming the current flowing is:

$$I = I_0 \sin \omega t$$

then the p.d. drop is:

$$V = |Z|I_0 \sin (\omega t + \arg Z)$$

or

$$V = ZI_0 \sin (\omega t + \delta)$$

1.6.3 The Complex Admittance

As $G = 1/R$ can be more useful than R, $1/Z$ can be more useful than Z.

The *complex admittance* is defined: $Y = 1/Z$

It has two parts, the conductance and the susceptance.

$$Y = G + jB$$

1.6.4 Power in A.C. Circuits Revisited

The instantaneous power dissipated by a circuit is given by:

$$P = IV$$

For $I = I_0 \sin \omega t$, $V = V_0 \sin (\omega t + \delta)$, the power dissipated by the component is:

$$P = I_0 \sin \omega t \, V_0 \sin (\omega t + \delta)$$

$$\Rightarrow \qquad P = V_0 I_0 \sin^2 \omega t \cos \delta + V_0 I_0 \sin \omega t \cos \omega t \sin \delta$$

Hence, the mean power over one cycle is:

$$<P> = \tfrac{1}{2}V_0I_0 \cos \delta \qquad (1.61)$$

or

$$<P> = V_{rms}I_{rms} \cos \delta \qquad (1.62)$$

The term $\cos \delta$ is called the power factor. If the load is purely reactive $\cos \delta = 0$, and if it is purely resistive $\cos \delta = 1$. Therefore, no power is dissipated in purely reactive loads. This causes great annoyance to the electricity board who pay a lot of money to push current through the grid to heavy industry, only to have most of the power wasted due to highly reactive loads such as machine-tool motors. Industrial power consumers are, therefore, charged at a rate which depends on their power factor, and often install large banks of capacitors outside the factory to bring their power factor back to 1.

If the waves are pure sinusoids then a shortcut can be taken.

$$<P> = \tfrac{1}{2} \operatorname{Re} \mathbf{V*I}$$

1.7 Analysis of Some Reactive Circuits

This section is an analysis of some common reactive circuits using the ideas developed in the previous sections.

1.7.1 The RC Integrator

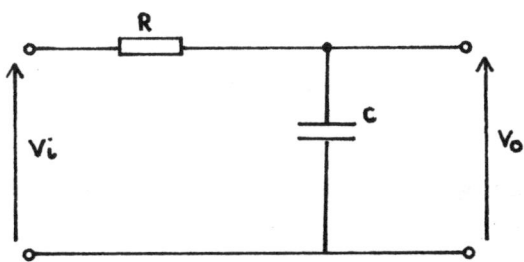

Figure 1.32 An RC integrator

Here we have:
$$I = C\frac{dV}{dt} = \frac{V_i - V_o}{R}$$

thus, if $V_o \ll V_i$
$$\frac{dV}{dt} \approx \frac{V_i}{RC}$$

$\Rightarrow$
$$RC\,V_o(t) = \int V_i\,dt \qquad (1.63)$$

So, provided that the input is applied for a small interval (so that V_o does not become large) the circuit integrates the input. This is useful in solving differential equations electronically, in signal averaging, sample & hold circuits and many other areas.

1.7.2 The RC Differentiator

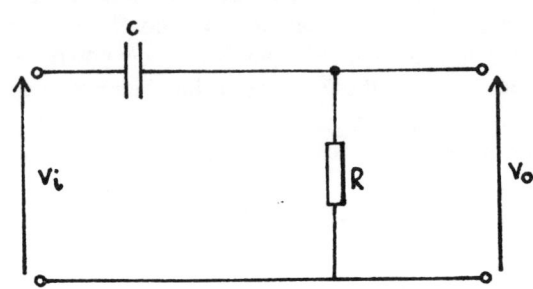

Figure 1.33 An RC differentiator

$$I = C \, d(V_i - V_o)/dt = V_o/R$$

if $dV_i/dt \gg dV_o/dt$ $\qquad V_o \approx RC \, dV_i/dt$ $\qquad\qquad$ (1.64)

a) $\qquad\qquad\qquad$ b) $\qquad\qquad\qquad$ c)

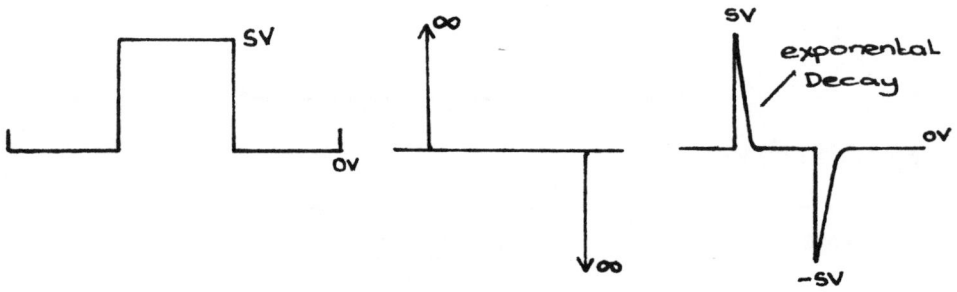

Figure 1.34 Differentiation of a square wave

The output is thus proportional to the rate of change of the input. This circuit is also useful in analogue computation (solving differential equations), but has many other uses. Suppose we wish to detect the 'edges' of a square wave input. If we apply a square wave input (a) to the circuit we should get output (b) but, in fact, we will get output (c). This can be modified by an amplifier to give a small pulse at the *leading* edge, and also a small pulse at the *trailing* edge if required.

1.7.3 Simple Passive Filters (1) RC Low Pass Filter

This circuit is essentially the same as the integrator, but here we consider what happens when we apply a sinusoidal signal.

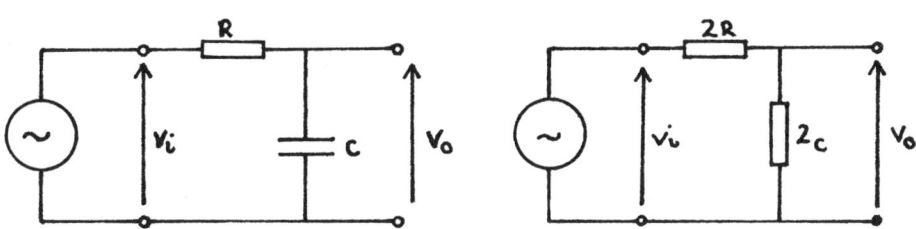

Figure 1.35 Analysis of an RC filter as a potential divider

We can think of the circuit as a potentiometer if we treat it as two impedances. Now, for a resistive *pot*, we can derive the *transfer function* which links the input to the output.

i.e. $$V_o = IR_2$$

where: $$I = V_i/(R_1 + R_2)$$

giving: $$V_o = V_i R_2/(R_1 + R_2) \quad \text{or} \quad V_o = TV_i$$

In the same way, the transfer function of the RC circuit can be calculated.

$$T_{RC} = Z_C/(Z_R + Z_C)$$

$$T_{RC} = 1/(1 + j\omega RC) \tag{1.65}$$

The action of the circuit is to *attenuate* (i.e. diminish in amplitude) signals with high frequencies, and not to affect signals with low frequencies. This is why it is called a filter, because it filters out signals with high frequencies.

Equation (1.65) is quite clumsy to use, and the information it carries can be expressed in a simpler form. At low frequencies we note that $T_{RC} \approx 1$, and at high frequencies that $T_{RC} \approx 1/\omega_{RC}$. .

At the intersection: $\qquad\qquad\qquad\qquad 1 = 1/\omega RC$

therefore, the frequency of the intersection (called the corner frequency, for obvious reasons) is:

$$f_C = 1/2\pi RC \qquad\qquad\qquad (1.66)$$

1.7.4 Simple Passive Filters (2) RC High Pass Filter

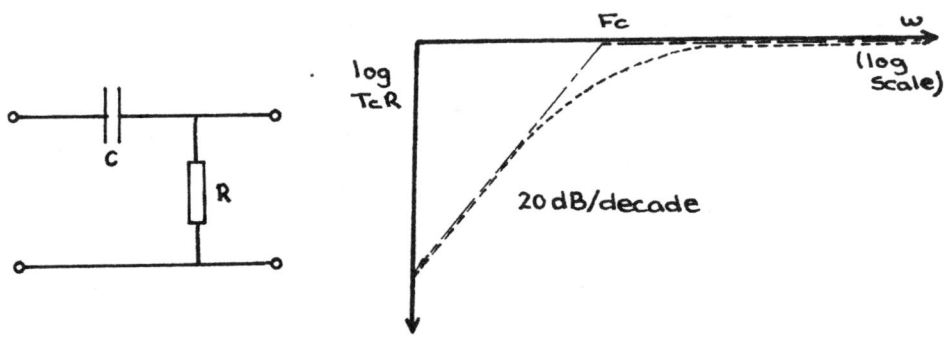

Figure 1.36 A CR filter and its transfer function

In the same way as before, the transfer function of the network can be found.

$$T_{CR} = R/\{R + (1/j\omega C)\} \qquad\qquad (1.67)$$

$\Rightarrow \qquad\qquad\qquad T_{CR} = j\omega RC/(j\omega RC + 1)$

Making approximations: $\qquad T_{f > fc} \approx 1$
$$T_{f < fc} \approx \omega RC$$

The corner frequency is still given by (1.66), even though the circuit is changed.

1.7.5 Series LCR Circuits

We wish to find the impedance of the series LCR circuit.

$$Z = R + j(\omega L - 1/\omega C)$$

And so:
$$Z = \sqrt{\{R^2 + (\omega L - 1/\omega C)^2\}} \qquad (1.68)$$

$$\delta = \tan^{-1}(\omega L - 1/\omega C)/R \qquad (1.69)$$

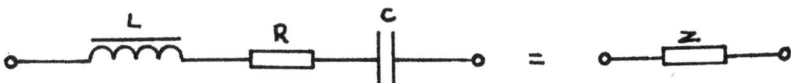

Figure 1.37 A series of LCR network

If we plot these we find that Z reaches a minimum, and the phase changes from $-\frac{1}{2}\pi$ at low frequencies (capacitive nature) to $\frac{1}{2}\pi$ at high frequencies (inductive nature). These effects are easy to understand, since at low frequencies $Z_L \approx 0$, and so the inductor can be effectively ignored, leaving a resistor and capacitor, and at high frequencies $Z_C \approx 0$ and the capacitor can be ignored, leaving a resistor and an inductor.

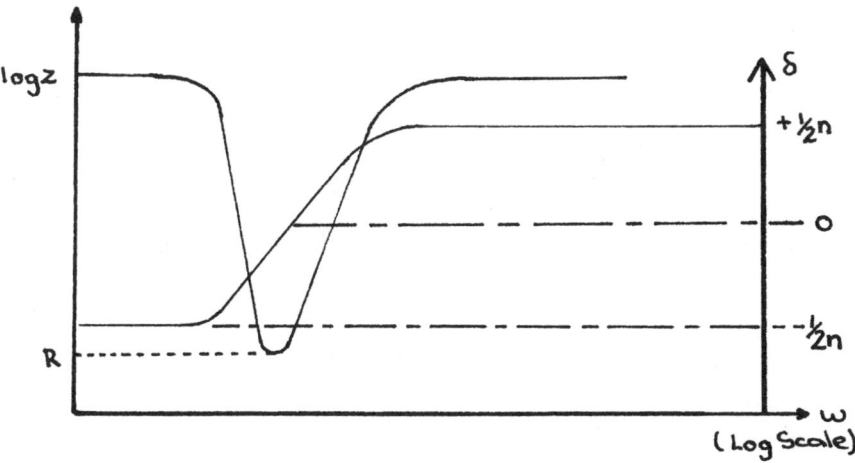

Figure 1.38 Bode plot of the series LCR network

The minimum can be found by calculus, but it is easier to do this by inspection. The minimum value is R, at which $\omega L = 1/\omega C$ and $\delta = 0$.

$$\Rightarrow \qquad Z_{min} = R \quad \text{at} \quad f = 1/2\pi\sqrt{(LC)} \qquad (1.70)$$

In analysis of LCR circuits the term 1/LC occurs quite regularly, and it is often referred to by the symbol:

$$\omega_0^2 = 1/LC \qquad (1.71)$$

This circuit has a high impedance to most frequencies, but at those around a special value (called the resonant frequency, since at this frequency current oscillations reach a maximum amplitude) it has a much lower impedance, and thus can be used to selectively filter out all frequencies except those around that value. In such a use it is termed a notch filter.

1.7.6 The LCR Parallel Resonant Circuit

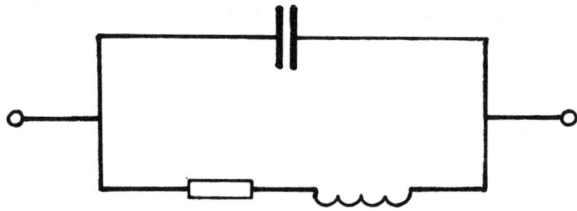

Figure 1.39 Parallel LCR Circuit

$$\mathbf{Z} = (R + j\omega L) \,\|\, 1/j\omega C$$

Note: the notation $A \| B$ means A in parallel with B, and for impedances $A \| B = AB/(A + B)$

$$\Rightarrow \qquad Z = \frac{(R + j\omega L)/j\omega C}{R + j\,(\omega L - 1/\omega C)}$$

This is an unpleasant expression, and the analysis of the circuit can be simplified if the complex admittance is used instead of complex impedance.

$$\mathbf{Y} = j\omega C + 1/(R + j\omega L) \qquad (1.72)$$

$$\Rightarrow \qquad \mathbf{Y} = \{(1 - \omega^2 LC) + j\omega RC\}/(R + j\omega L)$$

$$\Rightarrow \qquad Y^2 = \frac{\omega^2 R^2 C^2 + (\omega^2 LC - 1)^2}{R^2 + \omega^2 L^2} \qquad (1.73)$$

At voltage resonance Y is a minimum. The calculation to find when $\partial Y/\partial \omega = 0$ is tedious, and it is common practice to use *phase resonance* (i.e. when $\delta = 0$) as the principal condition for resonance for such circuits. This requires only that we find when Y is a real quantity.

from (1.7.16) $\qquad\qquad$ Im $Y = \omega C - \omega L/(R^2 + \omega^2 L^2)$

$\Rightarrow$ $\qquad\qquad\qquad\qquad\qquad \omega_p^2 = \omega_0^2 - R^2/L^2$ $\qquad\qquad\qquad$ (1.74)

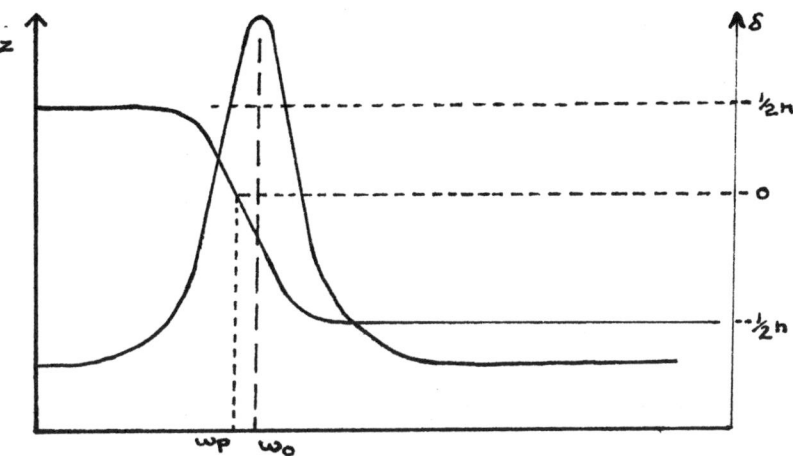

Figure 1.40 Bode plot of the parallel LCR network

For high Q circuits $\omega_p \approx \omega_0$. At phase resonance the circuit has impedance L/RC. This circuit has many uses. It is used in oscillators, since it can execute oscillations and the damping can be eliminated by amplification. It is used as a load in radio frequency amplifiers, since at resonance it need be driven by a very small current (as $Z_{parallel}$ is large) but a proportionately very large current flows around the circuit because *around the loop* it is a series circuit (as Z_{series} is small). This large current can be detected by using the primary of a transformer as the L in the LCR circuit, for example. Such amplifiers have high gains to signals at the resonant frequency ω_0, and low gains to all others. Thus, they can be used to select one radio station from the many being transmitted. (This is why the circuit is also known as a tuned circuit).

Questions for Chapter 1

1.1   Find the equivalent resistances of the following networks:

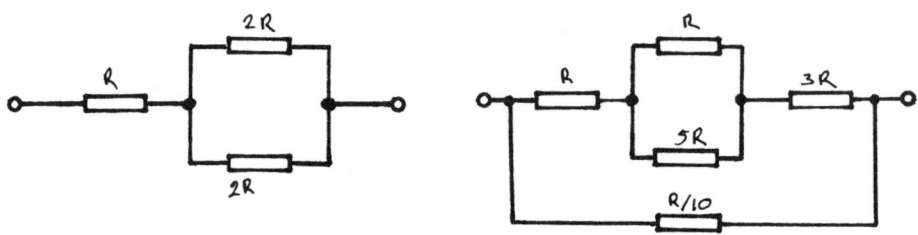

Figure 1.i Some resistor networks

1.2   I have an ammeter with F.S.D. 100 μA and internal resistance 1 275 Ω. Design circuits to use the ammeter as:

i   A voltmeter with F.S.D. 2 V.

ii   A voltmeter with F.S.D. 8.75 V.

iii   An ammeter with F.S.D. 20 mA.

iv   An ammeter with F.S.D. 1 A.

(hint: for voltmeters use the method in the text; for ammeters connect a resistance in parallel with the meter)

1.3   Find the voltages/currents measured, using the meters from 1.2, when the meters are driven from a source of e.m.f. 1.5 V and output resistance 600 Ω.

1.4   Find the current flowing in each resistor when 250 mA flows into a 12 kΩ and 22 kΩ resistor connected in parallel.

1.5   r, s, and t are three resistors connected into a Δ-network. Find ρ, σ and τ , which are three resistors connected into a Y-network, such that the resistance between any two Y-terminals is equal to that between the equivalent Δ-terminals. The remaining terminal will be left free.

(hint: equate the resistances between pairs of terminals in each network, and then solve the resulting equations for ρ, σ and τ in terms of r, s and t.)

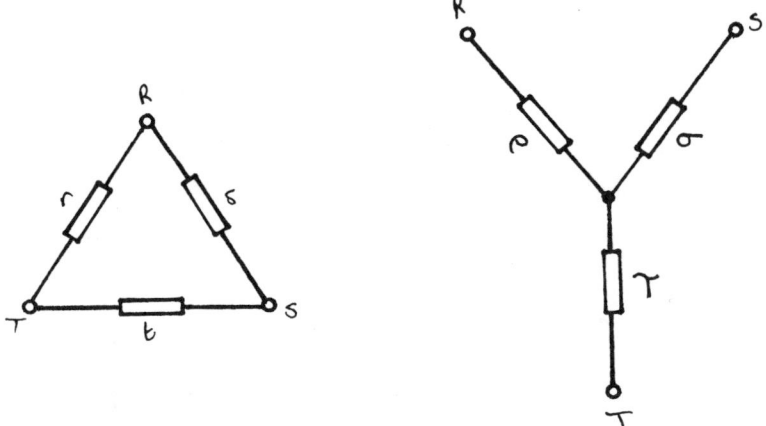

Figure 1.ii The delta and Y networks

1.6   Find the Thévinin equivalent circuits for the following networks:

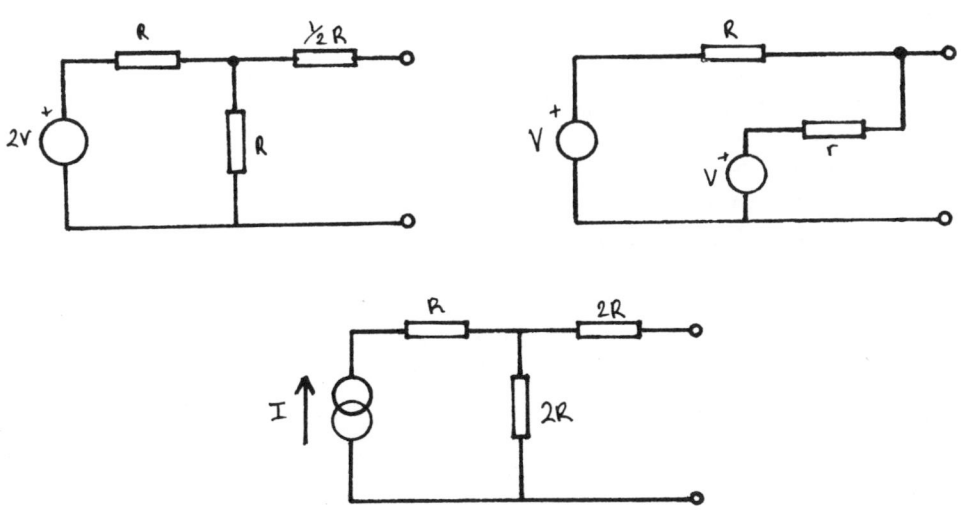

Figure 1.iii Some e.m.f, resistor, current source networks

1.7   Find the Norton equivalents for the networks of 1.6.

41

1.8   Find the maximum voltage, current, and power transferred to a load by the first
       network of question 1.6. Take R to be 100 Ω and V to be 5 V.

1.9   Find the power transferred to a 25 Ω load from a source of resistance 68 Ω and
       e.m.f. V. Take V to be 10 V d.c.; 14.1 V (peak) a.c.; 10 V (r.m.s.) a.c.

1.10   How long does the voltage across the capacitor, in figure 1.iv, take to rise to 5
       V, if it is initially discharged when the switch is closed?

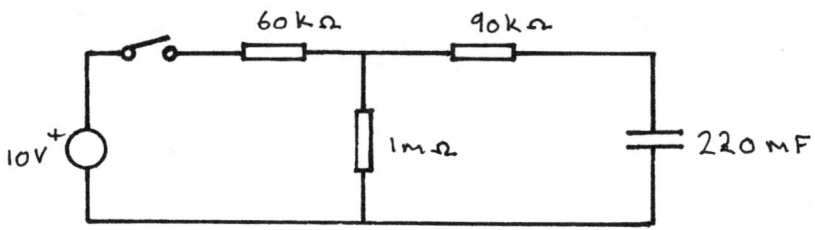

Figure 1.iv Charging a capacitor via a complicated network

Answers to Numerical Problems

1.1   2R; 0.098R
1.2   use resistors: 18.7 kΩ; 86.2 Ω; 6.41 Ω; 7.84 S (= 1/7.84 Ω)
1.3   1.46 V; 1.49 V; 2.47 mA; 2.50 mA
1.4   160 mA; 88 mA
1.8   5 V; 50 mA; 62.5 mW
1.9   290 mW; 290 mW; 290 mW
1.10   24.5 s

2

SEMICONDUCTOR DEVICES

2.1 Solid State Conductors

This section introduces a theory of the solid state which enables the behaviour of semiconductors to be explained.

2.1.1 A Revised Theory of Solid State Conduction

In §1.3 a theory of the structure of metals was introduced to help explain the existence of the *capacitor effect*. Although this theory is good for metals, it does not explain conduction in non-metals such as semiconductors, which is an effect that we have to consider. The new model retains the main features of the old model, but introduces additional details.

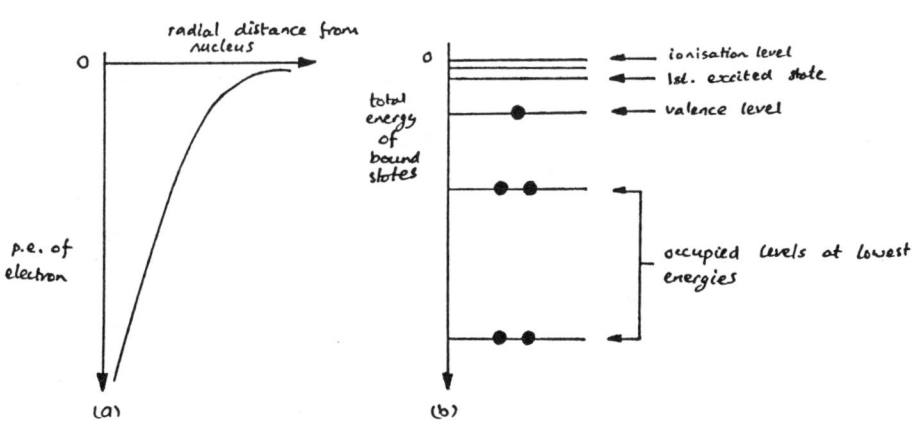

Figure 2.1 Allowed electron energy levels of an atom according to quantum mechanics

If we take a nucleus and plot the potential energy, of an electron, around the nucleus we get diagram (a). Using quantum mechanics, it can be shown that it is possible for an electron to exist in a number of discrete bound states. Each state in (b) corresponds to a particular electron energy.

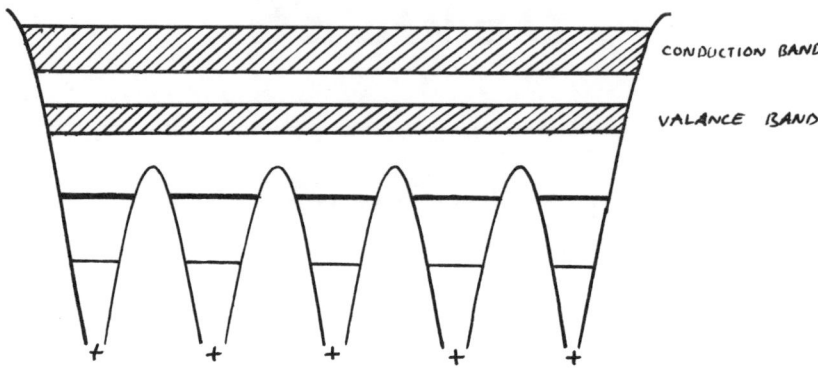

CONDUCTION BAND

VALANCE BAND

Figure 2.2 Energy bands in a periodic potential (crystal lattice)

In a crystal a great many atoms are close together, and the energy levels form into energy bands (due to the splitting of levels in adjacent atoms).

The lower band in Figure 2.2 is called the *valence band*, since it is occupied by the outermost electrons *(the valence electrons)* of the crystal, and the upper band, which corresponds to the first excitation level of a free atom, is called the *conduction band*. They are separated by an *energy gap*.

If the energy gap is large, and the valence band is filled, then the valence electrons cannot be accelerated by an external field (this requires that they gain extra energy, and they can only do this by moving into a higher level). Although the electrons are free to move about the crystal they cannot interact with an external field to produce an overall *drift* of electrons through the crystal. Such a crystal is an insulator. If the valence band is only partly filled then electrons can easily acquire energy from an external field, since there are many free levels. Such a crystal is a conductor, of which copper is a typical example. If the energy gap is small, but the valence band is filled, then the material will have a conductivity which is intermediate between that of insulators and conductors. Such a material is called an *intrinsic semiconductor*.

In a semiconductor the energy gap is small enough so that electrons can acquire sufficient energy from thermal excitations to cross into the conduction band. This has two effects. It increases the conductivity, since the electron density in the conduction band is increased, and the electrons leave *holes* in the valence band, which can be filled by nearby electrons. If a hole is filled by an electron a new hole is formed, and the effect appears to be the motion of holes, not of electrons.

44

Generally, the holes move randomly, but if a field is applied across the material the holes are found to act as if they were positive charge carriers, and acquire a drift velocity equal in magnitude, but opposite in direction, to the electrons. The conductivity of such an intrinsic semiconductor is very low when compared to that of a metal.

If we introduce, for example, a single atom of arsenic (As), which has five valence electrons, into a crystal of pure germanium (Ge), which is a semiconductor with four valence electrons, we find that only four electrons are used in bonds between atoms, and the remaining electron is weakly held to the atom. In fact, it is so weakly held that it can easily enter the conduction band and be carried away. This leaves the As nucleus with a *surplus* charge and releases an extra free electron. Arsenic is called a *donor impurity* because it has donated an electron to the conduction band.

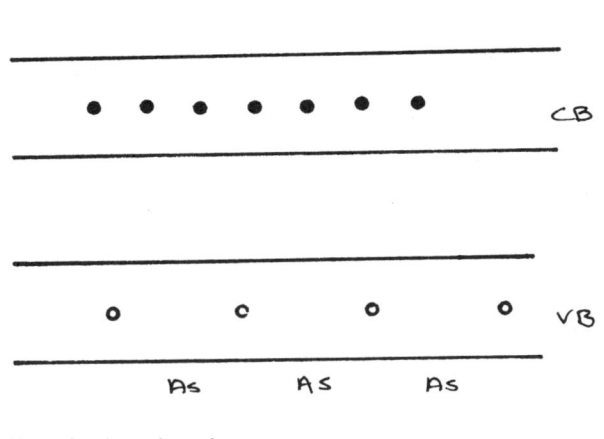

Figure 2.3 The effect of a donor impurity

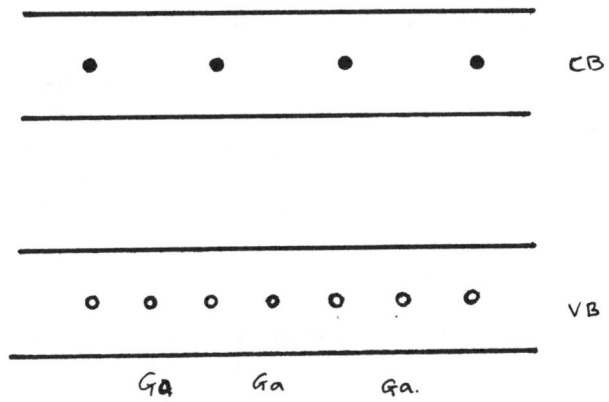

Figure 2.4 The effect of an acceptor impurity

A similar effect occurs when an atom of gallium (Ga) is introduced, except for the fact that the Ga atom has only three valence electrons, so it accepts an electron from the valence band, which is equivalent to donating a hole. It is called an *acceptor impurity*.

Both types of impurity drastically increase the conductivity of the semiconductor. Such materials are called *extrinsic semiconductors*, and with donor impurities are called n-type, since the majority of the conduction effects are due to the negative electrons, which are called *majority carriers*. With acceptor impurities the positive holes are the majority carriers and the crystal is called a p-type conductor. The process of introducing impurities is called *doping*.

2.2 PN Junction Diodes

This section introduces the *pn-Junction Diode*, a very important device. For completeness, and so that good circuits can be designed, I shall use the *diode-law*:

$$I = I_0 e^{qV/kT}$$

but I shall also deal with a small signal model, and a simple idealised model, which often provide great simplification. The last sections will deal with some of the special diodes which are available. Here the symbol, q, is used to represent the elementary (electronic) charge e $\approx 1.6 \times 10^{-19}$ C. This is to prevent confusion with the mathematical constant e ≈ 2.7182818. T = absolute temperature $\approx 273.16 + $ temp$/°$C. k = Boltzmann's constant $\approx 1.38 \times 10^{-23}$ J K^{-1}.

2.2.1 The pn-Junction

A *diode* is formed by doping an intrinsic semiconductor, usually silicon, with acceptor impurities, and then redoping part of the n-type region formed with donor impurities to make it p-type. A pn-junction is then formed in the material. When the junction is formed some electrons and holes diffuse across the junction and annihilate each other. The process continues until the resulting charge build up causes the creation of a potential barrier sufficient to stop further charge migration. A region is formed, at the junction, which has very few free holes and electrons and is, therefore, an insulator. This is called the *depletion layer*.

There are two ways in which electrons can pass the barrier from the low energy side to the high energy side. If they have sufficient kinetic energy they pass the barrier with only a loss of energy (just as a ball is slowed when it rolls up a hill); if they do not, they will be reflected by the barrier. There is also a quantum mechanical effect called *tunnelling* in which it is possible for a few electrons to pass the barrier, even though they do not have enough kinetic energy to *roll over* it. Electrons on the high energy side can *fall down* the barrier.

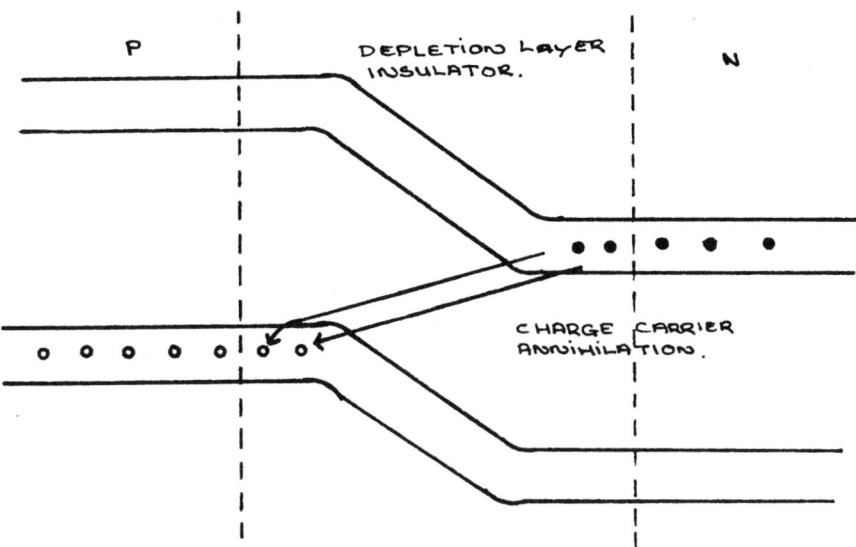

Figure 2.5 Simplified Model of band structure in a pn-junction diode

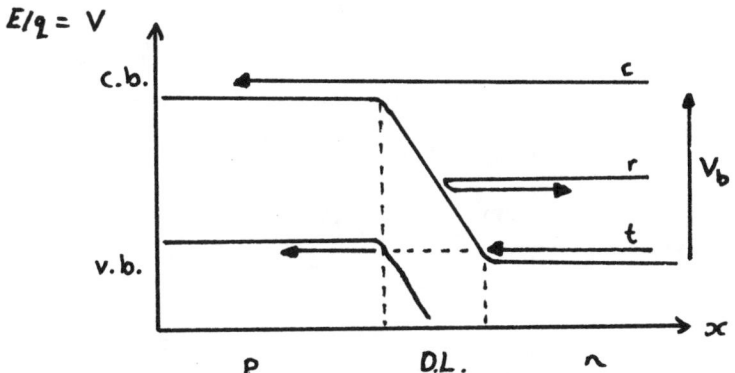

Figure 2.6 Electrons crossing (c), being reflected by (r) and TUNNELLING through (t) a potential barrier (V_b)

When an electron passes the barrier from the n-type side to the p-type side the most likely thing to occur is the combination of the electron with a hole, resulting in the destruction of both. This is called the *annihilation current*, I_A. When an electron falls down the barrier it generates a new electron in the n-type side, and this is called the *generation current* I_G.

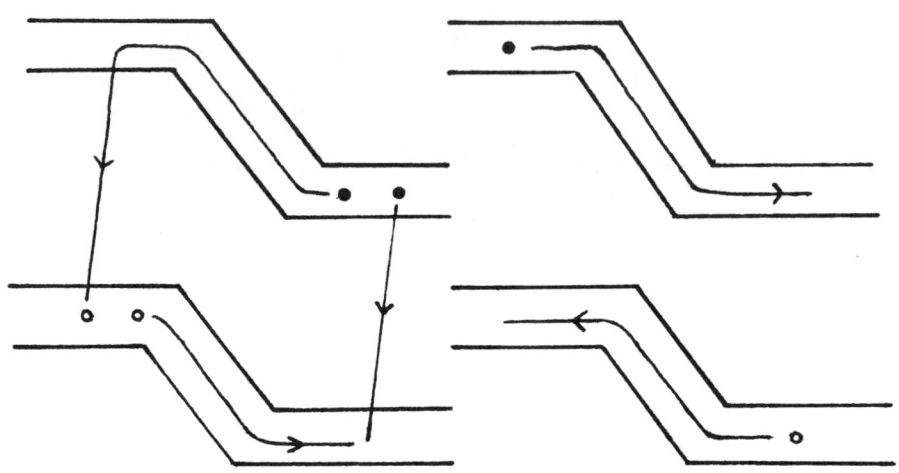

Figure 2.7 Annihilation and generation currents

The same things occur with holes, but in the other direction. Because the hole current is in the reverse direction to the electron current it has the same sign.

So:
$$i_A = i_{AE} + i_{AH}$$

and
$$i_G = i_{GE} + i_{GH}$$

With no external p.d. the system must be in equilibrium, otherwise it would act as a current source. i.e.

$$I = i_{A0} + i_G = 0 \tag{2.1}$$

The equilibrium can be disturbed by *forward biasing* the junction. This means applying a positive p.d. between the p-type region (the anode) and the n-type region (the cathode) and has the effect of increasing i_A to I_A ($I_A \gg i_A$), but does not affect i_G (the rate at which balls fall down a cliff does not depend on the height of the cliff).
in forward bias:

$$I = I_A + i_G \approx I_A \gg 0 \tag{2.2}$$

If, however, a negative p.d. is applied, the junction is said to be *reverse biased*. i_A is decreased and becomes approximately zero but, again, i_G is not affected.

in reverse bias

$$I \approx i_G = - i_{A0} \approx 0 \qquad (2.3)$$

This discussion can be made more qualitative by putting the device in the circuit of figure. 2.8.

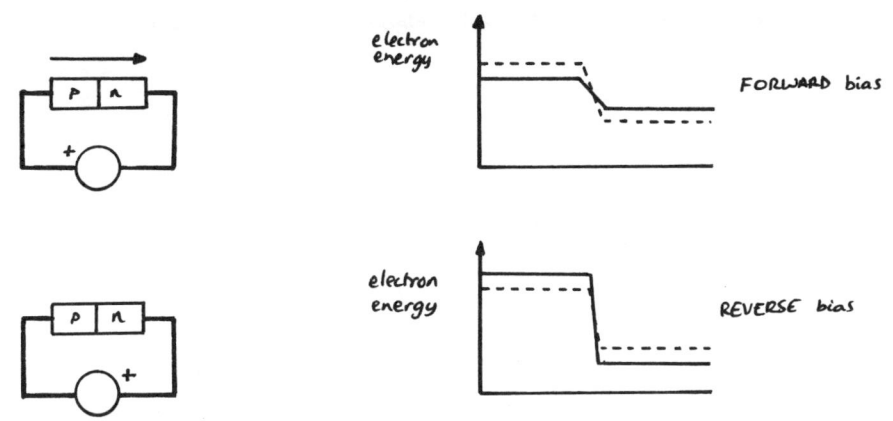

Figure 2.8 Effect of forward and reverse bias on barrier height

The *height* of the potential barrier is:

$$V_B = V_0 - V \qquad (2.4)$$

V_0 is the height of the barrier when $V = 0$.

It is found that:

$$i_A = I_1 e^{-qV_B/kT} \qquad (2.5)$$

This is based on statistical mechanics and I cannot prove it at this level. Physicists should recognise the *Boltzmann factor* $e^{\varepsilon/kT}$ (ε = energy = qV), which gives the number of electrons with sufficient energy to climb the barrier.

Substituting (2.4) into (2.5) gives:

$$i_A = I_1 e^{qV/kT} e^{-qV_0/kT}$$

If $V = 0$ then:

$$i_A = i_{A0} = - i_G = I_1 e^{-qV_0/kT} \qquad (2.6)$$

Writing $I_0 = - i_G$ and substituting (2.6) in (2.1):

$\Rightarrow$

$$I = I_0 e^{qV/kT} - I_0 \qquad (***)$$

or

$$I = I_0(e^{qV/kT} - 1) \qquad (2.7)$$

This is the diode law, and in most cases we neglect the 1 (for silicon $I_0 \approx 10^{-13}$ A, and for germanium $I_0 \approx 10^{-7}$ A) giving

$$I = I_0 e^{qV/kT} \tag{2.8}$$

which is the most useful form of the equation. Over a wide range of currents $V \approx 0.7$ volts for silicon diodes, and $V \approx 0.2$ volts for germanium. You can satisfy yourself that this is true by simple experimentation; alternatively, look at figure 2.12.

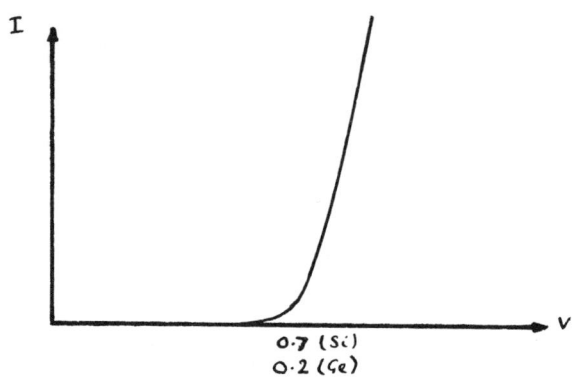

Figure 2.9 The diode curve

The asterisks (***) identify one of the faults in the derivation. I have identified an exponential factor with a constant, which is a fair approximation here, but not generally true!

The correct equation is (2.9)

$$I = AGT^{5/2} e^{-qV_{gor}/kT}(e^{qV/kT} - 1) \tag{2.9}$$

V_{gor} (the extrapolated bandgap voltage) is my constant V_0, and $I_1 = AGT^{5/2}$. A is the junction area, and G is a constant.

Equation (2.9) is rarely of use. It tells us that to make a power diode we require a larger junction area, and you will find that high current diodes are larger than small-signal diodes. It also allows us to find the temperature dependence of the devices.

$$(\partial I/\partial T)_V = \{2.5 + q(V - V_{gor})/kT\}I/T$$

$$\Rightarrow \qquad dI/I \approx 9\%/K$$

The general rule, for germanium and silicon devices, is that I doubles for every 8°C rise in temperature at room temperature and at constant V. Room temperature is assumed to be 17°C (290 K), and this allows a commonly made substitution

$$q/kT \approx 40 \quad \text{at room temperature}$$

2.2.2 The Simple Diode Model

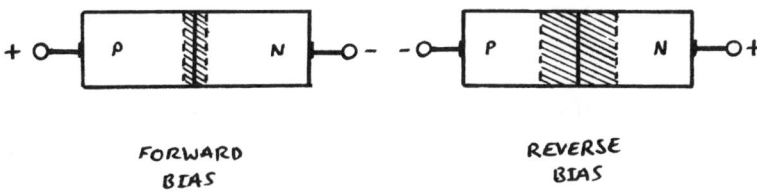

FORWARD
BIAS

REVERSE
BIAS

Figure 2.10 The depletion layer in forward and reverse bias

The previous model is correct in form. This is a model which is very useful. We noted before that the diode (with no external p.d.) has a depletion layer in which the charge carrier density is very low, and so acts as an insulator. The diagrams of figure 2.10 show how the p.d. applied across the diode changes the width of the depletion layer, and therefore alters the conductivity.

Reverse biasing increases the depletion layer width, so the device remains an insulator. Forward biasing decreases the width of the depletion layer, and so increases the conductivity. This model assumes that when the voltage across the diode reaches a certain limit called the *forward voltage drop*, V_F, the depletion layer is eliminated and the device becomes a conductor. When conducting, it drops V_F.

Forward Voltage Drops of Some diodes

germanium diodes	$V_F = 0.2$ V	
silicon	0.6 - 0.7 V	for signal diodes
	1.0 V	for rectifier (power) diodes conducting high currents
LEDs	2.2 V	(light emitting diode)

This simplifies circuit design, e.g. driving an LED.

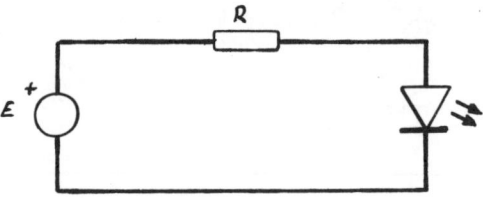

Figure 2.11 LED driver circuit

Calculation procedure

The maximum current will be determined by the diode (look in a data book or catalogue) or the circuit. For an LED it is approximately 30 mA (determined by the maximum power dissipation of the package).

$$I = I_{max}$$

If we are driving it from a voltage generator of e.m.f. E

from Ohm's law: $\qquad\qquad IR = E - V_F$

$\Rightarrow \qquad\qquad\qquad R = (E - V_F)/I \qquad\qquad\qquad (2.10)$

To drive a LED from a 9 V cell at 30 mA (high brightness):

$\Rightarrow \qquad\qquad\qquad R = (9 - 2.2)/0.03 = 226 \ \Omega$

Pick the nearest preferred value to this, which is 220R. It is really this simple; build one and see!

2.2.3 A Small Signal Diode Model

The diode-law, or the simple model, are useful when dealing with large signals. But what if we have the situation where we have a large constant signal with a small signal superposed (added) onto it? Because impedance is a linear operator then, if we apply superposed stimuli, the response is the sum of the responses to the individual stimuli.

if: $V' = V + v$

then $ZI' = V + v$ (I' = total current)

$\Rightarrow$ $I' = V/Z + v/Z = I + i$

where I = current if stimulus V is applied alone, and i = current if v is applied alone.

But as the diode is not a linear device the stimuli do not superpose in this manner.

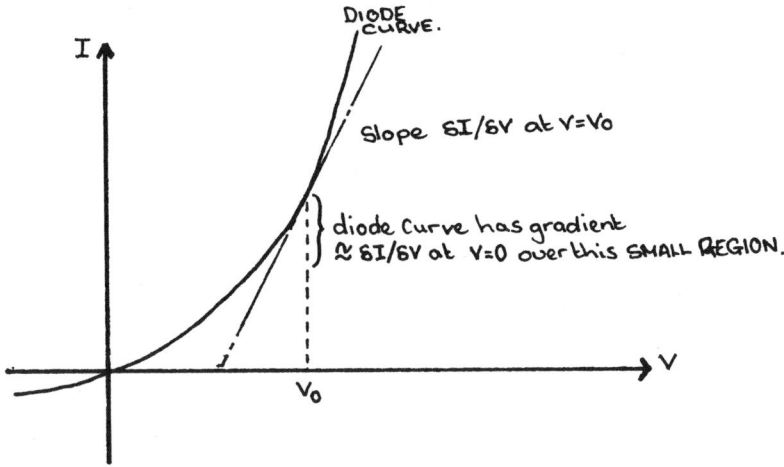

Figure 2.12 Construction to find the small signal resistance of a diode

---

So, it looks like we are in for terrible mathematics if we apply a stimulus which is a large constant signal with a small variable signal superposed upon it. However, if the signal, v, is sufficiently small the gradient of the diode curve is approximately constant. We assume that it is constant, and take the reciprocal gradient $\delta V/\delta I$ to be the dynamic resistance of the diode, denoted by r (do not confuse this with internal resistance, which can also be denoted by r).

By doing this we can work out the response to the large signal, I, using the diode law (unfortunate, but necessary) and the small signal response from the dynamic resistance. We then superpose these to get a (usually good) approximation to the actual response. In this situation a small signal is one of a few millivolts, whereas a large signal is one of a few volts.

What is r?

$$I = I_0 e^{qV/kT}$$

$$(\partial I/\partial V)_T = qI/kT \implies (\delta I/\delta V)_T \approx qI/kT$$

giving:
$$g \approx 40I \tag{2.11}$$

(g is the conductance $\delta I/\delta V = 1/r$)

or
$$r = 0.025/I \tag{2.12}$$

Remember this as 25 ohms per milliamp of large signal current.

If the signal v is small enough that we may write $v = \delta V$ and $i = \delta I$ then

$$i \approx v/r$$

∴
$$I' \approx I_0 e^{qV/kT} + vg \tag{2.13}$$

or
$$I' = I_0(1 + qv/kT)e^{qV/kT} \tag{2.14}$$

Calculation Procedure

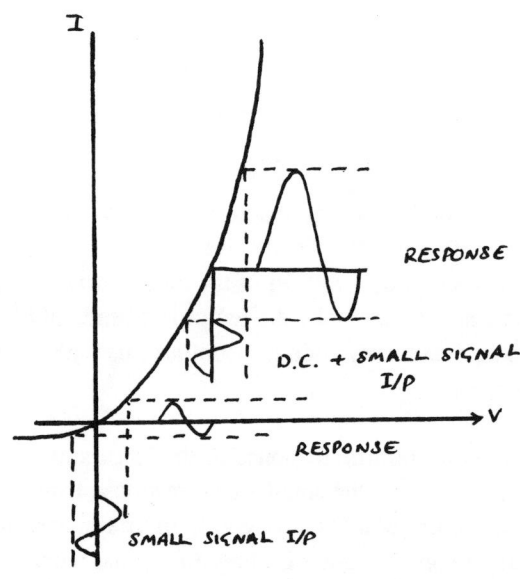

Figure 2.13 Response of a diode to superimposed large and small signals

54

1   work out large signal response I from diode law;
2   work out the conductance, g, from I;
3   find small signal response from i = vg; and
4   add these to find the total current I + i.

If the small stimulus, v, becomes too large you will find distortion introduced, due to the non-linearity of the diode.

2.2.4 A Final Point, Reverse Breakdown

The diodes, as I have shown, essentially conduct in one direction only. It is a bit unrealistic to suppose that if we put an infinite reverse voltage across the diode it would conduct zero current. That situation is impossible, but it is possible to generate a potential of several million volts in the laboratory, using Van der Graaf generators. The force on an electron due to high potentials is large enough to rip it out of the atom it is bound to and so, in a diode, large reverse currents flow when the potential exceeds a critical value, called the peak inverse voltage (PIV). When this occurs, large currents begin to flow in the device and when the heat being produced exceeds the amount which can be lost through conduction, radiation, etc. the device gets very hot, and is eventually destroyed. The full diode curve is shown in figure 2.14.

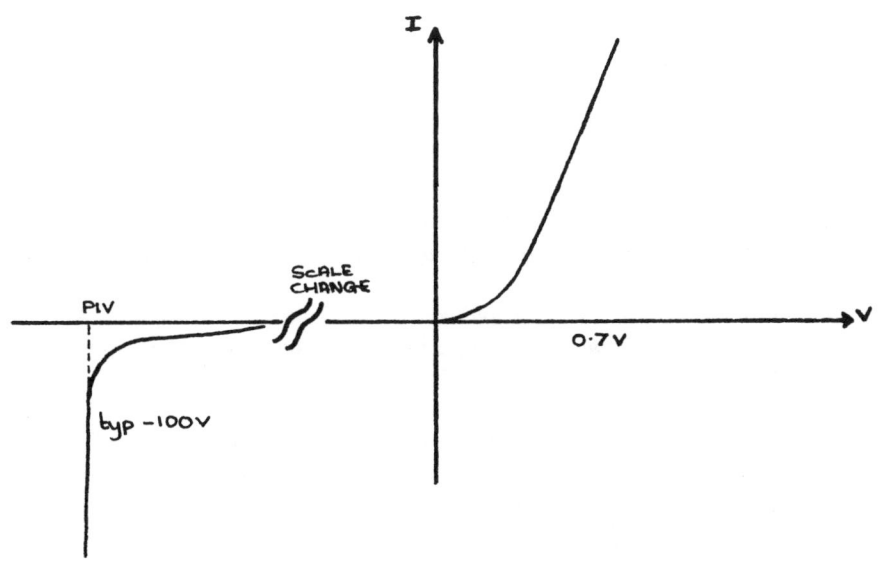

Figure 2.14 Full diode curve

The PIV can range from a few volts to a few thousand volts depending on the device.

2.2.5 Diode Applications - Rectifiers

Now we know how diodes work - but what do we use them for? We have 240 V a.c. delivered to our houses, but most electronic circuits work on approximately 3 - 30 V d.c. Buying batteries is very costly but, by transforming the 240 V a.c. to a low voltage and then changing to d.c. using a diode circuit, the need for batteries can be eliminated.

Simple PSU Rectifier

As an example, I shall design a circuit to provide d.c. for the simple diode circuit described earlier. The first stage is to *step down* the 240 V a.c. to 6 V a.c. This is done with a transformer, as explained in §1.5. The a.c. must be changed into d.c. Diodes conduct in one direction only so by connecting a diode in series with a load, across the transformer, only the positive part of the a.c. is applied to the load.

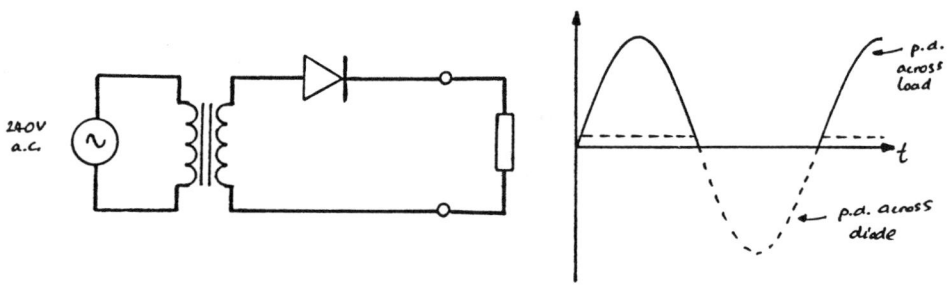

Figure 2.15 Simple half wave rectifier

This circuit is, in fact, good enough to just drive a LED, but let us be a bit more ambitious and design a circuit which can be used to replace a PP3 cell in most cases.

From an examination of the circuit waveforms it is clear that the circuit draws current during half of each cycle. It would be better to take current during the entire cycle. The circuit of figure 2.20 is called a *bridge rectifier*.

If $V_{AB} > 0$ then diodes 1 & 3 are forward biased and current flows through the load from the top of the circuit to the bottom. If $V_{AB} < 0$ then diodes 2 & 4 are forward biased and current flows through the load from top to bottom. Therefore, the current through the load only flows in one direction - it is d.c.

The output is a *'lumpy'* d.c., which is better, but still not much like that obtained from a cell. Lumpy d.c. can be thought of as a steady d.c. level with an a.c. component added to it. If you connect a large capacitor across the output it will act almost as a short to the a.c. component, but also as a very high impedance to the d.c. component. Thus the d.c. component will be dropped across the capacitor, but not the a.c. component, and so the supply taken from the capacitor is d.c.

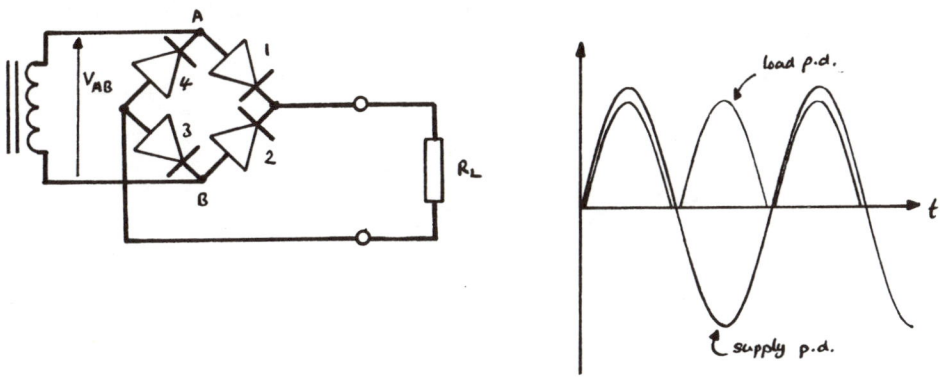

Figure 2.16 Bridge rectifier

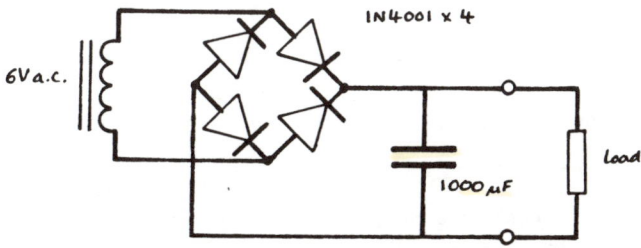

Figure 2.17 Smoothed bridge rectifier

Connecting the capacitor (called the *smoothing capacitor*) does not, however, give perfect d.c. It has a *ripple* superposed to it instead of a large a.c. component.

More Detailed Analysis of P.S.U. Circuit

The output from the transformer is V volts a.c. rms. This means that the peak input to the bridge rectifier is $V\sqrt{2}$. The peak output of the bridge is thus:

$$V_{BOP} = V\sqrt{2} - 2V_F \qquad (2.15)$$

When the bridge output voltage is greater than the p.d. across the capacitor, charge flows onto the capacitor. When it is less, the capacitor discharges through the load (it cannot discharge through the bridge). Since there is very little resistance in series with the capacitor, during charge up V_C follows V_{BO}, but the discharge is through the larger load resistance and so the amount of ripple present depends on the resistance of the load. If a very large capacitor is used (such as 1000 μF - 10 000 μF) then the time constant (RC) is very large, over a wide range of loads, and the ripple is small. The output of the circuit is approximately V_{BOP}.

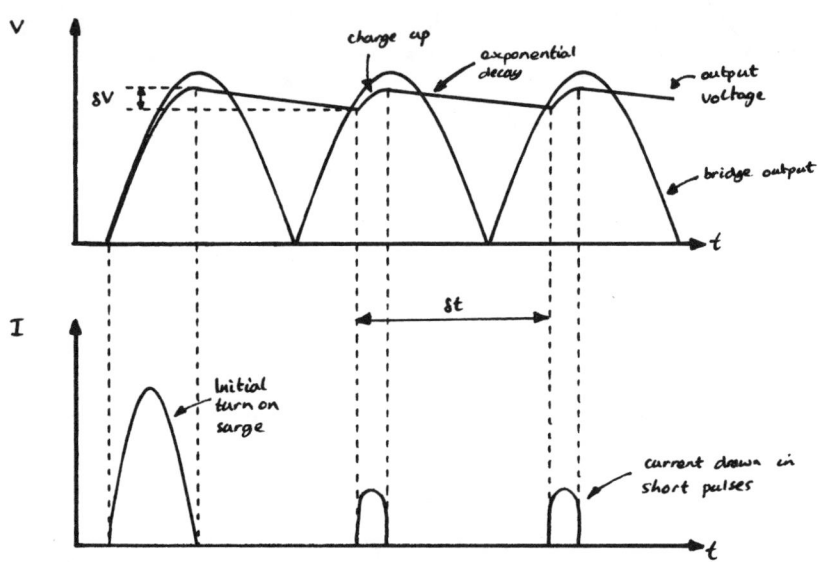

Figure 2.18 Waveforms of bridge rectifier

Because the current flows into the capacitor only when the voltage across the capacitor, V_C, is less than the bridge output, the current is taken from the supply in short pulses. This means that the power (energy per unit time) dissipated during the pulses is greater than if a smaller current was drawn continuously.

It was stated above that the output voltage of the circuit is approximately V_{BOP}. However, if the ripple has peak-to-peak value δV then the mean output voltage is going to be:

$$V_o \approx V_{BOP} - \tfrac{1}{2}\delta V$$

∴
$$\delta V \approx - V_{BOP}\delta t/R_L C \qquad (2.16)$$

For (2.16) to be valid $R_L C \gg \delta t$

$$\Rightarrow \qquad V_o \approx V_{BOP}(1 - \delta t/2R_L C)$$

substitute in (2.15) $\qquad V_o = (V\sqrt{2} - 2V_F)(1 - \delta t/2R_L C) \qquad (2.17)$

Equation (2.17) allows the smoothing capacitor and transformer, to give a required output voltage and ripple, to be chosen. In Britain we have 50 Hz mains and so $\delta t \approx 10$ ms (USA ≈ 8 ms).

It is useful to express the ripple as a fraction of the output voltage. This is ripple factor.
$$k_r = V_{Rrms}/V_o \qquad (2.18)$$

V_{Rrms} is the r.m.s. amplitude of the ripple, i.e. $\frac{1}{2}\delta V/\sqrt{2}$

using (2.16) & (2.17) $\qquad k_r \approx 1/\{(R_L C/\delta t - 1)\sqrt{8}\} \qquad (2.19)$

Assuming that the load resistance $\gg 100$ ohms for the diode P.S.U. then:

$$k_r \ll 0.04 \quad \Rightarrow \quad V_{Rrms} \ll 0.3 \text{ V} \quad \Rightarrow \quad V_o \gg 6.8 \text{ V}$$

If $R_L C \gg \delta t$, as is required for validity of equation (2.16), then the '1' in (2.19) may be dropped.

$$\therefore \qquad k_r \approx 1/f_r R_L C\sqrt{8} \qquad (2.20)$$

This is the commonly used definition of the ripple factor. (f_r is the ripple frequency = $1/\delta t = 100$ Hz in the UK and 120 Hz in the USA)

It is possible to reduce the ripple to microvolts using active circuits (regulators), as discussed in a later section. The circuits also reduce the internal resistance of the P.S.U., so that the output voltage does not change detectably over large current ranges (e.g. microamps to amps, a change of 120 dB).

2.2.6 Other Diodes

Zener Diodes

A zener diode is a special diode which is very useful in clamps or P.S.U.s. It is constructed so that the current in reverse breakdown is limited and so the device is not damaged in reverse breakdown. Because of the nature of the breakdown characteristic the diode can conduct a large range of currents, in reverse breakdown, without the voltage drop changing much from the P.I.V. Zeners are manufactured with low P.I.V.s (volts, not hundreds of volts) and are always connected in reverse.

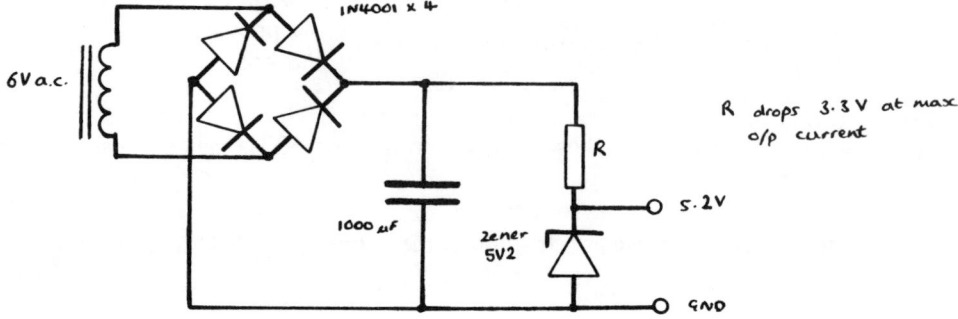

Figure 2.19 Zener diode regulator

Light Emitting Diodes

As an electron is annihilated (by filling a hole) it releases energy. This energy comes out as light (a photon). By picking the correct semiconductor materials the light will be visible light (commonly red, orange, green, or yellow). Infra-red LEDs can also be made.

A variation on this theme is the semiconductor laser.

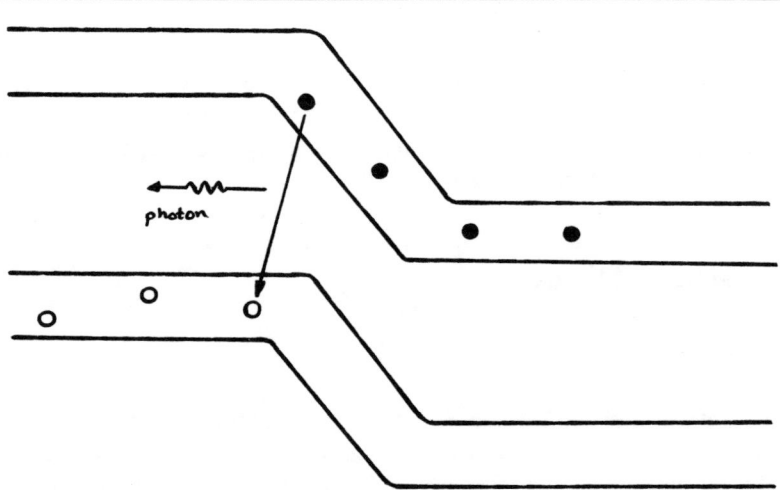

Figure 2.20 Light emission at a pn-junction

Photodiodes

As well as light emission (which occurs in LEDs) light absorption can occur. This is when an electron in the valence band absorbs the energy of a photon and jumps into the conduction band, causing the creation of an electron and hole pair and, therefore, increasing the current. These optoelectronic devices will be studied in greater detail in chapter 6.

Tunnel Diodes

Tunnel (Esaki) diodes are devices made with an ultrathin depletion layer and have an interesting characteristic curve.

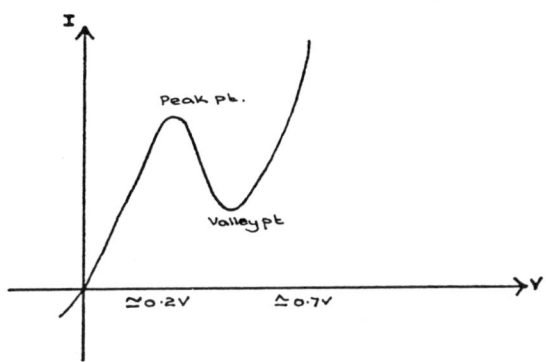

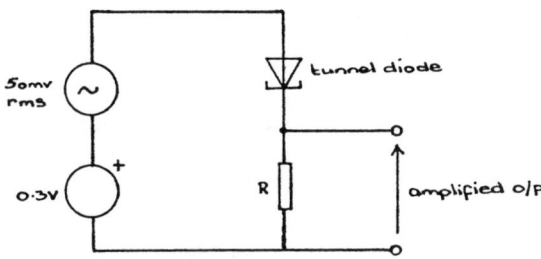

Figure 2.21 Tunnel diode curve and application

In the valley region the dynamic resistance, r, is negative. When set up as a potentiometer, and biased into the valley, the device can act as a small signal amplifier.

The small signal transfer function for such a circuit is:

$$t = R/(R + r)$$

In the valley region $R + r < R$, since $r < 0$, and so $t > 1$ and the circuit amplifies instead of attenuating.

In the 1950s these devices were hailed as the solution to a great many circuit problems because they were fast and required only simple circuits. However, they turned out to be hard to use, and transistor technology has rapidly advanced, so they are not now in general use.

Varactor (Varicap) Diodes

When a diode becomes reverse biased, electrons are drawn out of the p-type area, leaving it slightly positively charged. To balance this electrons flow into the n-type area, from the circuit, until the n-type area is slightly negatively charged.

What do we have then? We have an area with a positive charge, an insulating area, and an area with a negative charge. A capacitor!

When a diode is brought rapidly out of reverse bias this capacitive effect shows itself in what is called the *reverse recovery time*.

Most diodes are manufactured to reduce this capacitance to a minimum, but varactor diodes are manufactured to deliberately exploit the capacitance.

It is found that the capacitance depends on the reverse bias voltage, so the device is an electronically programmable capacitor. They are used in modern tuners to allow electronic (i.e. digital and *intelligent*) tuning of radios. For example, a circuit could be designed to make the tuner seek out the strongest signal near a particular frequency, which would compensate for drift in the transmitter, or to sweep across the waveband until it found a signal, etc.

Schottky Diodes

In high speed switching circuits the reverse recovery time is a problem. The Schottky (or hot carrier) diode has a special construction and features a low forward voltage drop ($V_F \approx 0.3$ V) and a very fast switching action.

The device exploits the fact that aluminium, used to make contacts in integrated circuits, can act as an acceptor impurity when in contact with silicon. This problem is prevented in circuits by forming a very heavily doped n-type region (written n^+, n = ordinarily doped, n^- = lightly doped) where a n-type region is connected to a contact.

In the Schottky diode this is not done on one of the contacts, making a slight barrier between semiconductor and metal. The resulting device is a Schottky diode.

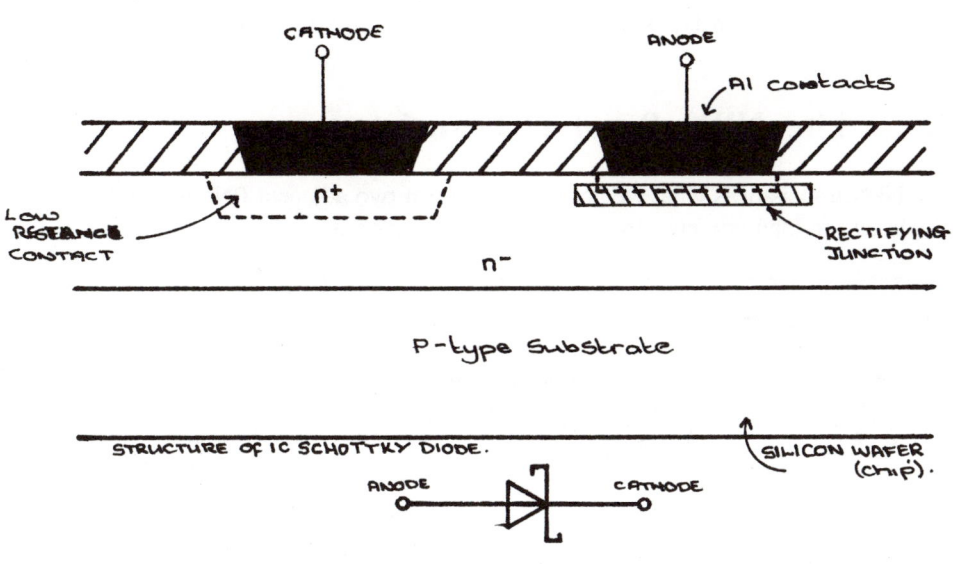

Figure 2.22 Structure of an IC Schottky diode

2.3 Bipolar Transistors

This section introduces the bipolar (or diffusion) transistor. They are essentially very simple devices, and most readers with some experience of electronics will probably have some experience of designing circuits around the *current amplifier model* of the transistor.

That is the model in which base current and collector current are related by the factor β (h_{FE}):

$$I_C = \beta I_B$$

If so, you may be quite surprised to learn that designs based on this equation are, in fact, bad designs, because the parameter β can vary quite considerably in a sample of *identical* transistors. For example, the cheap transistor ZTX300 has $50 < \beta < 300$. How can you design an amplifier around this?

I shall use the more accurate Ebers-Moll model:

$$I_C = I_{C0} e^{qV_{BE}/kT}$$

63

Not only is this equation more accurate, but it explains how we can build amplifiers with no base resistors (such as differential amplifiers and operational amplifiers) and allows us to derive a very useful small signal model.

Throughout this section I shall use the NPN transistor. All the equations can be converted to PNP transistors by changing the sign of both I_{C0} and V_{BE}

2.3.1 The NPN Bipolar Transistor

The NPN transistor is what we get when we form two adjacent PN junctions on a slice of silicon (germanium, etc). In fact we dope one region n^+.

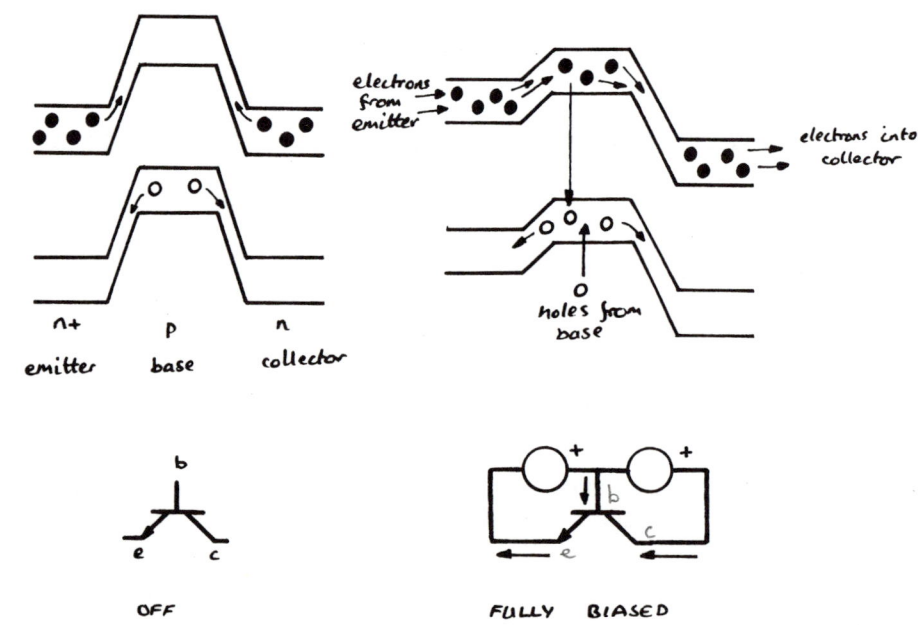

Figure 2.23 Simplified model of band structure in an npn transistor

The n+ region is called the *emitter* because it emits electrons. The n region is called the collector because it collects electrons. The p region is called the *base* because the transistors used to be formed with the p region as the base of the device (for PNP substitute holes for electrons above).

When the emitter-base junction is forward biased, and the collector-base junction reverse biased, large numbers of electrons enter the base from the emitter. Because the emitter is much more heavily doped than the base, most of these electrons do not combine with holes to form the annihilation current of the emitter-base diode. The excess electrons *diffuse* across the base region until they reach the collector. They then fall down the potential cliff formed by the reverse biased collector-base diode and form a large collector current.

It can be seen from figure 2.24 that the collector current is not dependent on either the collector-base voltage, V_{CB}, or the collector-emitter voltage, V_{CE}. It should only depend on the number of excess electrons entering the base and, therefore, on the base-emitter voltage, V_{BE}. In reality there is a very slight dependence, but significant errors are not introduced by ignoring this.

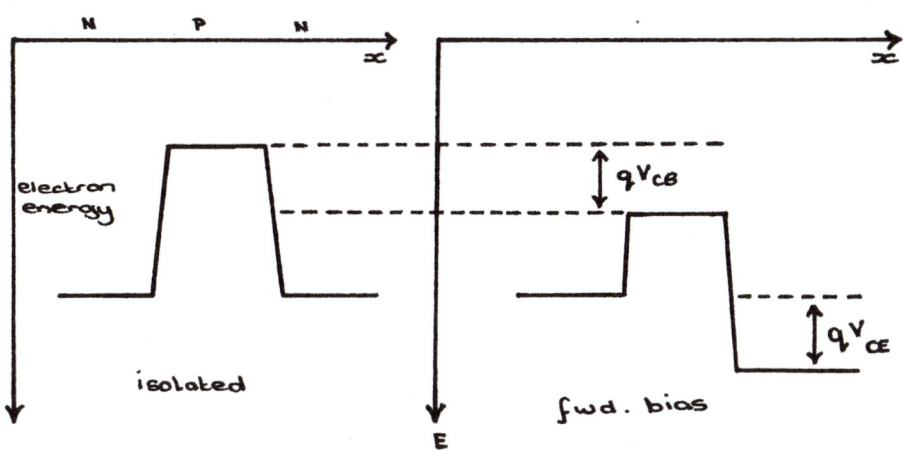

Figure 2.24 Electron potential energy in an npn transistor

Suppose that some electrons enter the base. The number which diffuse across into the collector depend on two numbers. These are the *emitter injection efficiency*, γ, and the *base transport factor*, δ. γ & δ are both properties of any particular junction, not *universal constants*.

$\therefore$
$$I_C = \delta\gamma I_E \qquad (2.21)$$

I_E is the emitter current.

This equation defines the constant α:

$$\alpha = I_C/I_E \qquad (2.22)$$

The base-emitter junction is *abrupt*, so it can be described using the diode law:

$$I_E \approx I_{E0}e^{qV_{BE}/kT}$$

$$\Rightarrow \qquad I_C = \alpha I_{E0}e^{qV_{BE}/kT}$$

$$\Rightarrow \qquad I_C = I_{C0}e^{qV_{BE}/kT} \qquad (2.23)$$

This is the Ebers-Moll model. Transistors are designed to make $\alpha \approx 1$, so I_{C0}(silicon) $\approx$ 10^{-13} A, etc. This means that the transistor will begin to conduct at around 0.7 V, as we assume in the simple model.

2.3.2 The Current Amplifier Model

How can we make this *diode like* equation give us the simple model based on the parameter β? Because the emitter-base junction is a diode, we can say that over almost all values of collector current $V_{BE} \approx 0.7$ V. In analogy to the simple diode model we can say that the transistor will not be *turned on* unless there is at least 0.6 V across the base (0.2 V for germanium transistors).

From Kirchoff's law: $\qquad\qquad I_B = I_E - I_C$

equation (2.2) $\Rightarrow \qquad\qquad I_B = (1 - \alpha)I_E$

$\Rightarrow \qquad\qquad\qquad I_C/I_B = \alpha/(1 - \alpha)$

or $\qquad\qquad I_C = \beta I_B$ with $\beta = \alpha/(1 - \alpha) \qquad (2.24)$

Although the device is a transconducting amplifier (I depends on V) it can be treated as a current amplifier!

$\alpha \approx 1$, but varies slightly from transistor to transistor. Consequently $1 - \alpha$ is a number ≈ 0 which varies slightly from transistor to transistor. This means that β is a large number which varies wildly, making a very badly defined parameter.

2.3.3 A Small Signal Transistor Model

A small signal transistor model is very useful. The small signal conductance is derived in a similar manner to that for the diode. Here is it called the *mutual conductance* because it relates the input to output.

$$g_m = (\delta I_C/\delta V_{BE})_T = i_C/v_{BE} \approx qI_C/kT = 40I_C \qquad (2.25)$$

The small signal behaviour of the transistor is essentially determined from two simple rules:

$$i_C = g_m / v_{BE} \qquad (2.26)$$

and

$$v_{CE} \text{ is not a function of } i_C$$

This means, to an external circuit, the collector appears to be connected to a current *sink* of magnitude i_C.

Also:

$$i_B = i_C / \beta = g_m / \beta v_{BE} \qquad (2.27)$$

This is what we would get if there was a resistor of magnitude g_m / β between base and emitter with the current source joined to the emitter or, alternatively, if there were a resistor of magnitude α / g_m between emitter and base, with the current source joined to the base. For small signals the transistor can be replaced with either one of these models.

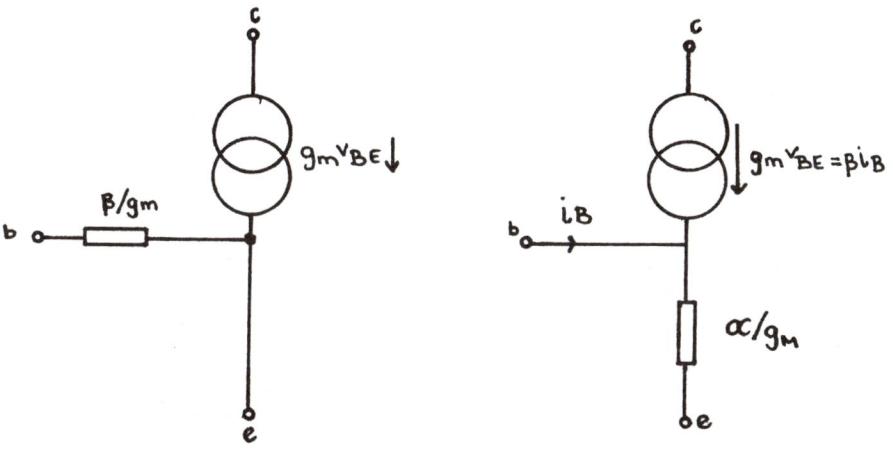

Figure 2.25 Small signal models of a bipolar transistor

2.3.4 Bipolar Transistor Amplifiers I - Emitter Follower

This circuit is useful because, although it has no voltage gain, it can be used to give a power gain (for example, in hi-fi outputs), to match high impedance outputs to low impedance inputs, etc.

The reason it is called an *emitter-follower* (or *common collector*) is because the output follows the voltage at the emitter (and the collector is common to both input and output circuits).

$$V_o = V_i - V_F \qquad (2.28)$$

For large signal inputs the transistor is on, and V_F is always somewhere between 0.6 V and 0.7 V. The emitter current is determined by the load resistance, R_E.

The emitter current is: $\qquad I_E = V_o/R_E = (\beta + 1)I_B \qquad (2.29)$

Neglecting $V_F \Rightarrow \qquad V_i/I_B = (\beta + 1)R_E \qquad (2.30)$

This is the large signal input resistance of the circuit, sometimes referred to as the resistance 'looking into the base'. The output resistance is found by 'looking into the emitter' and is, in fact, just R_S/β.

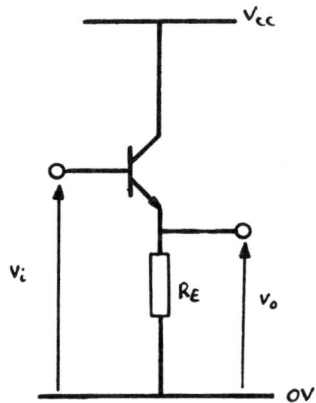

Figure 2.26 Simple emitter follower

The small signal behaviour of the circuit is similar, and is best determined using the 'α/gm' model. The results of such a calculation are quoted below:

$$v_o = R_Evi/(R_E + \alpha/gm) \approx 1 \qquad (2.31)$$

$$r_{in} = \beta/g_m + (\beta + 1)R_E \approx \beta R_E \qquad (2.32)$$

$$r_{out} = R_S/(\beta + 1) + \alpha/g_m \qquad (2.33)$$

The effect of the circuit is to make the emitter resistor, R_E, appear at the input to be a much larger resistor $\approx \beta R_E$. This is the whole point of an emitter follower.

Emitter Followers and Clipping Distortion

The output voltage can only take values between the supply voltage and earth. What happens when you drive the circuit into a region where the output should exceed these limits is called *clipping distortion*, see figure 2.29. This effect causes a deterioration of tonal quality if the signal is music.

Input Near Ground

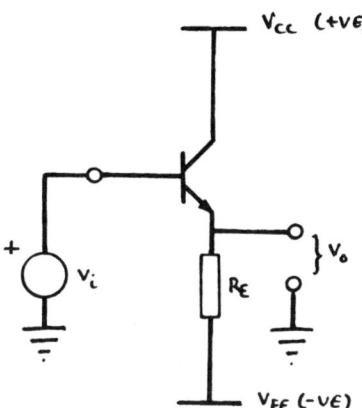

Figure 2.27 Emitter follower for use with a signal centred about zero volts

Clipping distortion can be combatted in a number of ways. If the input is centred at zero volts we can exploit this by replacing the supply voltages (V_{CC} & 0 V) with a symmetrical supply V_{CC} and V_{EE} (= - V_{CC}).

Biased Emitter Follower

Alternatively, we can force the base to rise to some positive potential. This process is called *biasing*. In order to prevent the biassing from being 'upset' by the input signal, d.c. inputs must be excluded. This is done by capacitively coupling (a.c. coupling) the signal input to the transistor base.

The potential the emitter sits at when there is no input is called the *working point*, (or quiescent point). The emitter current in such a state would be called the *quiescent emitter current*, and other currents/voltages are similarly named.

When an input is applied the output swings about the quiescent point, not about the mean of the input, which is usually 0 V. It is usual to choose the quiescent point to allow maximum symmetrical output swing, so it is put at approximately $\frac{1}{2}V_{CC}$.

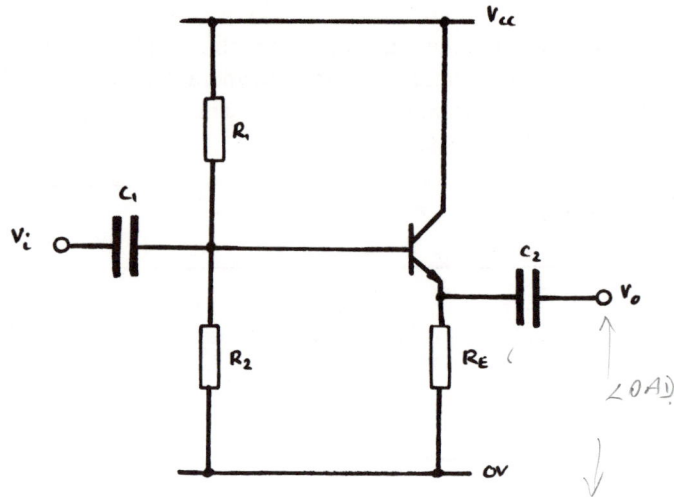

Figure 2.28 Biased, small signal, emitter follower

To ensure maximum transfer of voltage from the biasing circuit to the amplifier we should make the output resistance of the biasing circuit less than that looking into the base.

$$\text{i.e. } R_1 || R_2 \ll \beta R_E$$

The output resistance of bias circuit is $R_1 || R_2$ because the supply voltage is assumed to be an e.m.f. with zero resistance.

Example Design Procedure

An example circuit specification is that: i) the circuit should work over the audio range (20 Hz - 20 kHz); ii) the maximum output swing is to be 4V; and iii) the power dissipation of the transistor must not exceed 50 mW.

$2 \times 4 V = \left(8v\right) + \left(2v + 2v.\right) HEADROOM$

Choose V_{CC} to be 12 V as this allows room (2 V at top and bottom) between the allowed output extremes and the maximum and minimum possible outputs.

This means the working point, $V_{E0} = \frac{1}{2}V_{CC}$, is 6V.

In a worst case, the transistor could have 12 V between collector and emitter, so this requires $I_{E0} < P_{max}/V_{CC}$, i.e. $I_{E0} < 4.2$ mA. I shall choose $I_{E0} = 1$ mA, which requires that $R_E = 6$ kΩ, 6k2 is the nearest preferred value.

70

Next R_1 and R_2 must be chosen. $V_{B0} = V_{C0} + V_F$, so here $V_{B0} = 6.9$ V (corrected for the fact that V_{E0} is not quite 6 V), and so R_1 drops 5.1 V and R_2 drops 6.9 V

$$\Rightarrow \qquad R_1 : R_2 = V_{CC} - V_{B0} : V_{B0}$$

here: $\qquad R_1 : R_2 = 5.1 : 6.9 = 1 : 1.35$

This gives two conditions:

$$R_2/R_1 = 1.35$$
and $\qquad R_1| |R_2 \ll \beta R_E$

I can call the second condition: $\quad R_1| |R_2 \approx 0.1\beta R_E$

Using a ZTX300, $\beta_{min} = 50$ and so $R_1| |R_2 = 31$ kΩ. Solving these equations simultaneously gives $R_1 = 54$ kΩ and $R_2 = 73$ kΩ. The nearest preferred values are $R_1 = 56$k, $R_2 = 75$k.

The input capacitor C_1 forms a high pass filter with the load it sees, which is $(R_1| |R_2)| |\beta R_E$. The bias network dominates the input impedance and so the load seen by the capacitor is approximately $R_1| |R_2$, or 32 kΩ. I want to fix the corner frequency below 20 Hz. Using $f_C = 10$ Hz implies $C_1 \approx 500$ nF.

The output capacitor C_2 is a high pass filter in combination with an unknown load impedance, but it is safe to assume that the load impedance will be more than R_E. A minimum load resistance is, therefore, R_E, and so $C_2 \approx 2.6$ µF. There are now two cascaded filters, and so the attenuation at f_C will be 6 dB, not 3 dB. To be on the safe side, double the values of the capacitors. $C_1 = 1$ µF, $C_2 = 4.7$ µF.

A Few Notes on the Circuit

The major advantages of this circuit are the biasing and the a.c. coupling. The biasing means that the amplifier can handle signals which would not be sufficient to turn an unbiased transistor on. The a.c. coupling means that the output is centred about zero volts, and so the diode drop which occurs within the circuit is eliminated, and the output does exactly follow the input.

There are drawbacks. When no input is applied large currents still flow through the transistor, giving rise to power wastage, especially in very high power applications. *Class A operation.*

Other Voltage Followers

If you wish to amplify (give power gain to) small positive d.c. signals you cannot use the above circuit. For a signal which will never approach the positive supply voltage a d.c. coupled PNP emitter follower can be used. This will always be on when the input is near zero volts.

Alternatively, for high power output stages in which quiescent currents are not really wanted, a *complementary emitter follower* is suitable. This is a fusion of NPN and PNP emitter followers, which can source and sink currents. When the input is greater than 0.6 V the NPN transistor conducts, and the output follows the input. When the input is less than -0.6 V the PNP transistor conducts and the output follows the input.

This circuit suffers from the effect illustrated in figure 2.29, which is called *crossover distortion*. It gives drastic deterioration in the the quality of reproduced sound at low input levels. The solution is to bias the transistors just into conduction.

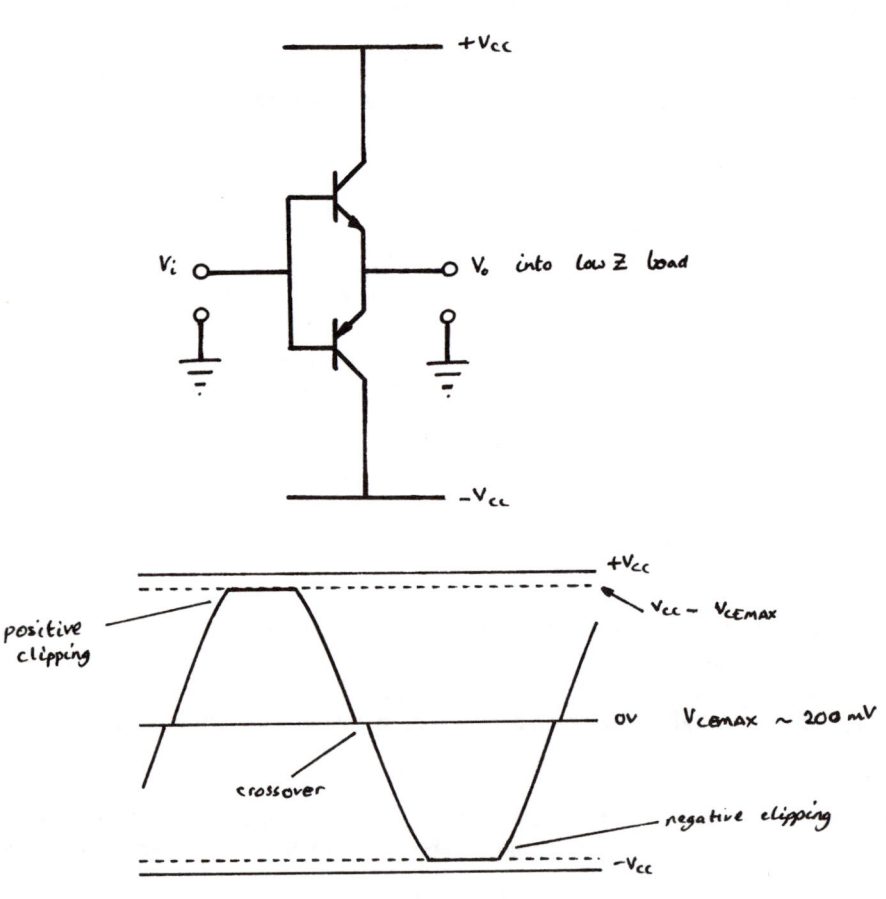

Figure 2.29 Complementary emitter follower

2.3.5 Bipolar Transistor Amplifiers II - Common Emitter

This amplifier is useful because not only does it give power gain, it also gives voltage gain. I shall consider it in two forms: a simple common emitter amplifier (rarely used), and the much more versatile and stable common-emitter amplifier with emitter degeneration.

Simple Common Emitter Amplifier

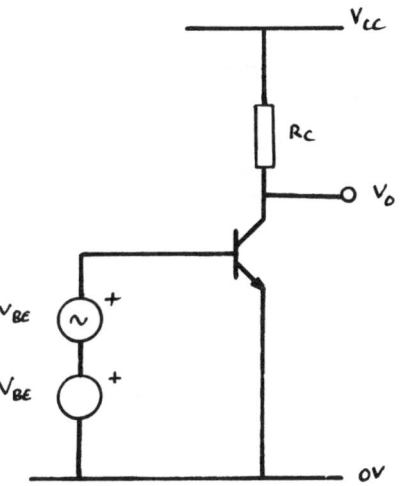

Figure 2.30 Common emitter amplifier

In the above, V_{BE} is the biasing input, and v_{BE} is the small signal input that we actually want to amplify.

The amplifier conducts a quiescent collector current, I_C, in response to the d.c. biasing input. The total output current is:

$$J_C = I_C + v_{BE}g_m \tag{2.34}$$

And so the output at the collector is:

$$V_C = V_{CC} - I_C R_C - v_{BE}g_m R_C \tag{2.35}$$

This is a small signal output, superposed on the quiescent output. So to small signals only:

$$v_o = - g_m R_C v_{BE} \qquad (2.36)$$

The small signal voltage gain of the circuit is v_o/v_i.

$$a_V = - g_m R_C \qquad (2.37)$$

This circuit can be used to give a very high voltage gain. Unfortunately, the quiescent current, and therefore the gain, is temperature dependent. The dependence of I on temperature, derived for a diode from equation (2.9), also applies for the transistor. Since $g_m = 40 I_C$, when the temperature rises by 8°C the gain is doubled. This behaviour cannot be permitted in any good design.

Common Emitter with Emitter Degeneration

The working of this circuit is analogous to the emitter follower, except for the fact that the output is taken above the collector (not below the emitter) to give an inversion. The circuit is biased to give a working point halfway between the supply terminals, i.e. $V_{C0} = \frac{1}{2} (V_{CC} + V_{EE})$, but the biasing network is not shown.

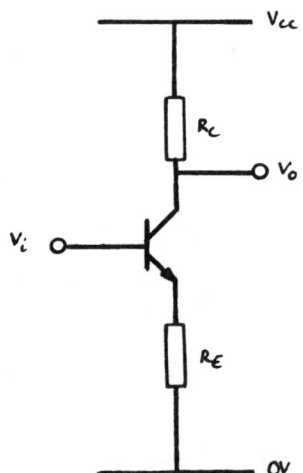

Figure 2.31 Common emitter with emitter degeneration

The biasing keeps the transistor well into the *active* region.

$$\therefore \qquad\qquad\qquad V_E = V_i - V_F \qquad\qquad\qquad (2.38)$$

and
$$\qquad\qquad I_E = V_E/R_E = (V_i - V_F)/R_E \qquad\qquad (2.39)$$

For real transistors, $\alpha \approx 1 \Rightarrow \qquad I_C \approx I_E$

$$\therefore \qquad\qquad I_C = V_i/R_E - V_F/R_E \qquad\qquad (2.40)$$

The voltage at the collector is $\qquad V_o = V_{CC} - I_C R_C$

equation $(2.40) \Rightarrow \qquad V_o = V_{CC} - V_i R_C/R_E - V_F R_C/R_E \qquad (2.41)$

Differentiating (2.41) wrt. V_i gives the small signal gain.

$$v_o/v_i = - R_C/R_E \qquad\qquad\qquad (2.42)$$

This gain is not a function of I_C, and so is temperature independent.

This circuit cannot be used in all circumstances. For the circuit to have a low output resistance R_C must be *small*, typically a few kilohms. R_E must be of the same order as R_C, so that the working point is well away from V_{EE} and the output can swing symmetrically towards V_{CC} or V_{EE}. Thus, the gain is nearly always less than -10.

This problem can be overcome by looking at temperature effects in a different way.

The full transistor law could be written:

$$I_C = f(T, V_{BE})$$

i.e. the collector current depends on both ambient temperature (in fact, junction temperature) and on base-emitter voltage.

Therefore, from the chain rule,

$$dI_C = (\partial I_C/\partial V_{BE})T \, dV_{BE} + (\partial I_C/\partial T)_V \, dT$$

or, in small signal notation:

$$i_C = g_m v_{BE} + h_T t$$

The parameter h_T is the rate of change of I_C with respect to junction temperature, at constant base-emitter voltage. (It is not one of the 'h-parameters'.)

What this equation is saying is that temperature can be considered just as an *input* to the transistor amplifier. Temperature changes always occur at low frequencies (certainly much lower than the bottom of the audio range) so, providing we only want to amplify small signals above about 10 Hz, steps can be taken to cut the gain at low frequencies.

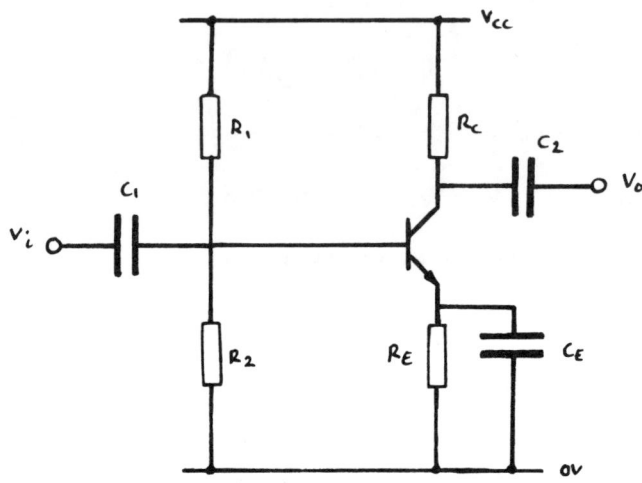

Figure 2.32 Common Emitter with Emitter Degeneration

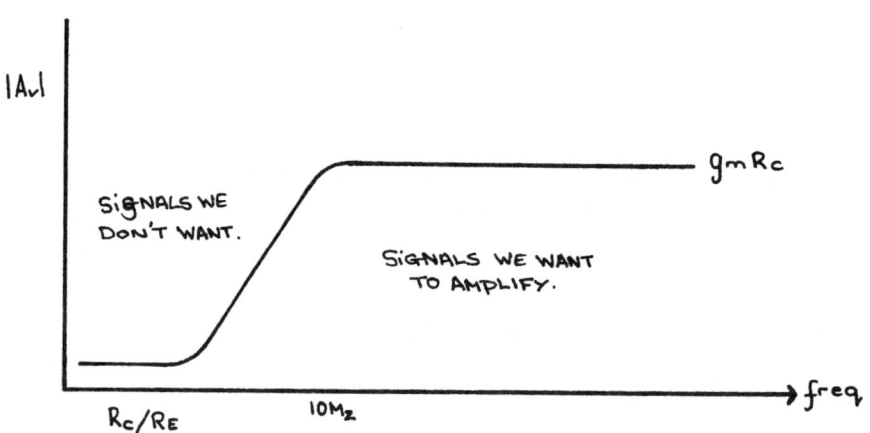

Figure 2.33 The Gain of the New Circuit

R_E & C_E are chosen such that, at frequencies above about 10 Hz, the capacitor acts to short the emitter resistor to ground, giving a high gain $\approx -g_m R_C$. At low frequencies $R_E \gg X_C$ and the gain is $\approx -R_C/R_E$. By choosing R_E to make this 1, or less, the effect of temperature changes are eliminated and a much more stable, high gain, a.c. amplifier results. The graph in figure 2.33 shows how the gain varies with signal frequency.

If a well defined gain, less in magnitude than $g_m R_L$, is required, only part of the emitter resistor need be bypassed.

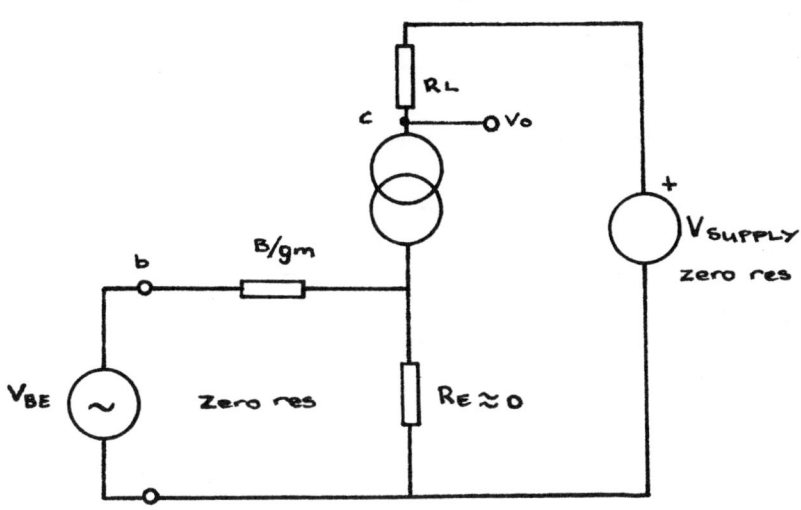

Figure 2.34 Small signal model of common emitter amplifier

Small Signal Input and Output Resistances of the Circuit

From the diagram: $\qquad\qquad r_{in} = \beta/gm$ $\qquad\qquad\qquad\qquad$ (2.43)

$$r_{out} = R_C \qquad\qquad\qquad\qquad (2.44)$$

2.3.6 Load Lines and Transistor Characteristics

The full transistor characteristic curve (the graph that represents the interrelation of variables - i.e. V_{BE}, I_C, V_{CE}) cannot be represented on graph paper. However, cross-sections of the three dimensional surface can be.

The curves are flat for $V_{CE} > V_{CEsat}$ since I_C does not depend on V_{CE} except for potentials below the limit. It is possible to do a graphical analysis of an amplifier using this curve, but this is not really as good as using the equations discussed above.

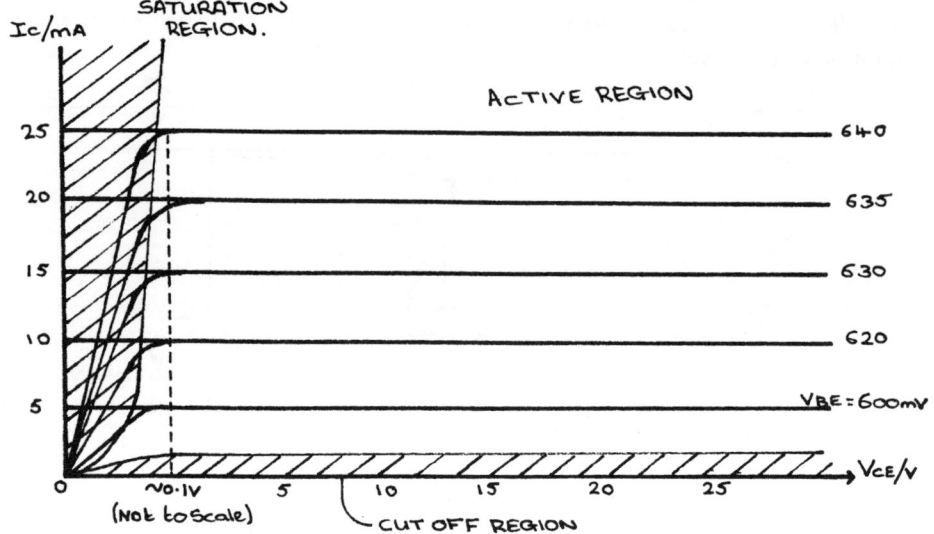

Figure 2.35 Characteristic of bipolar transistor

Method of Load Lines - for Common Emitter Amplifier

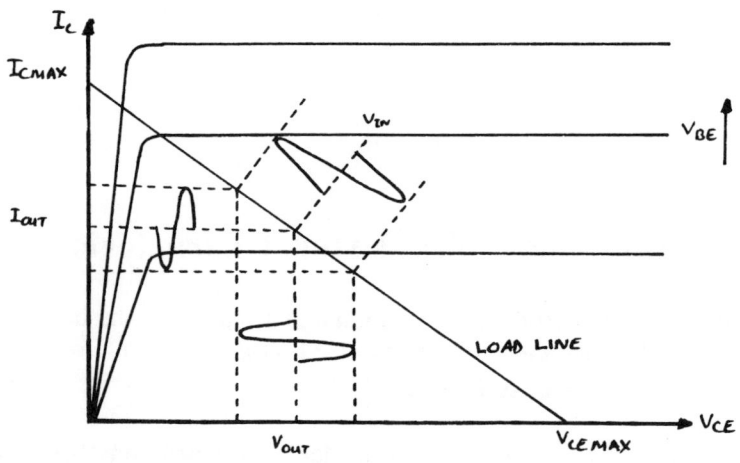

Figure 2.36 Analysis of amplifier using 'load lines'

1   Sketch the characteristic and circuit;

2   From the circuit, decide the maximum values of I_C, and V_{CE};

3   Plot these points on the characteristic, and join them with a line;

4   From the circuit, calculate the maximum output swing about the working point, and mark these points on the graph. This limits the line drawn in step (3);

5   You now have a rough transfer characteristic, which you can use to analyse inputs and outputs.

2.3.7 Bipolar Transistor Amplifiers III - The Long Tailed Pair - or Differential Amplifier

These names come about due to the appearance of the circuit, and its action, which is to respond to the differences in the inputs. Usually the output is taken from one collector but it can be taken as the difference in the outputs from both collectors).

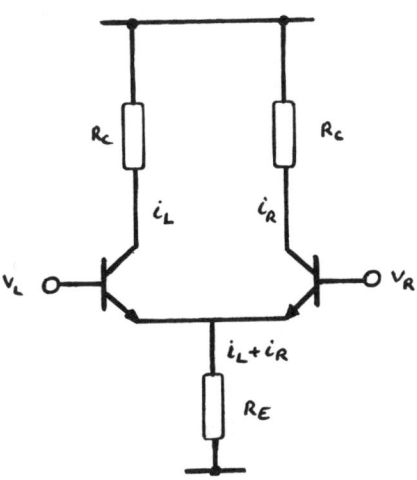

Figure 2.37 Amplifier used in analysis of L.T.P. amp.

The small signal analysis of the amplifier is relatively straightforward, but involves some unpleasant looking fractions. figure 2.37 shows a typical small signal differential amplifier. The inputs are d.c. centred about 0 V.

$$i_L = g_{mL} v_{BEL} \qquad (2.45)$$

From standard equations:

$$i_R = g_{mR} v_{BER} \qquad (2.46)$$

By choosing identical collector resistors the quiescent currents can be made to match. $\therefore g_{mL} = g_{mR} = g_m$. Also, if the base currents are insignificant, then $v_{BEL} = v_L - (i_L + i_R)R_E$ and $v_{BER} = v_R - (i_L + i_R)R_E$. Making these substitutions, and then adding equations gives:

$$i_L + i_R = g_m(v_L + v_R)/(1 + 2g_m R_E) \qquad (2.47)$$

and subtracting:

$$i_L - i_R = g_m(v_L - v_R) \qquad (2.48)$$

Finally, subtracting both these equations gives the collector current of the right hand side transistor, which is where the output is usually taken from.

$$i_R = - \tfrac{1}{2}g_m(v_L - v_R) + \tfrac{1}{2}g_m(v_L + v_R)/(1 + 2g_m R_E) \qquad (2.49)$$

$\Rightarrow$

$$v_o = \tfrac{1}{2}g_m R_C(v_L - v_R) - \tfrac{1}{2}g_m R_C(v_L + v_R)/(1 + 2g_m R_E) \qquad (2.50)$$

This can be written in the more compact form (2.51).

$$v_o = a_{DM}(v_+ - v_-) - a_{CM}\tfrac{1}{2}(v_+ + v_-) \qquad (2.51)$$

Differential Mode and Common Mode Gain

The reason for using this connection is its immunity to noise and temperature effects. The transistors are usually mounted close together, so they pick up the same noise and are at the same temperature. Suppose I wanted to amplify some d.c. signal v_{in}, but a single transistor amplifier picks up a noise signal, v_n, and suffers temperature changes equivalent to v_t. I can apply the signal to the *inverting input* v-, i.e. v_R, with the *non-inverting input* tied to ground.

i.e.

$$v_- = v_{in} + v_n + v_t$$

$$v_+ = v_n + v_t \qquad (2.52)$$

The difference between v_+ and v_- is referred to as a *differential mode* input, whereas the signal which is part of both v_+ and v_- is called a *common mode* input. The symbols a_{DM} and a_{CM} represent the differential mode and common mode gains.

$$v_o = - a_{DM} v_{in} - a_{CM}(\tfrac{1}{2}v_{in} + v_n + v_t) \qquad (2.53)$$

Equation (2.53) gives the output of the amplifier when the signals (2.52) are applied. The circuit responds to both the differential mode and common mode inputs but, because the differential mode gain is typically much larger than the common mode gain, the response to common mode signals can be neglected.

i.e.

$$v_o \approx a_{DM}(v_+ - v_-) \qquad (2.54)$$

Common Mode Rejection Ratio

The common mode rejection ratio, or CMRR, is the ratio of differential gain to common mode gain.

$$CMRR = a_{DM}/a_{CM} \qquad (2.55)$$

This can be simply calculated from (2.51).

i.e.
$$CMRR = \tfrac{1}{2}g_m R_C / \{ g_m R_C / (1 + 2g_m R_E) \}$$

$$\Rightarrow \qquad CMRR = \tfrac{1}{2} + g_m R_E \approx \tfrac{1}{2} + 20 V_{EE} \qquad (2.56)$$

For an amplifier with a ±15 V power supply the CMRR is 50 dB, and so the response to common mode signals is insignificant.

Differential Output

Equation (2.51) is the output taken, at the right hand side collector, but a similar expression is found for the output at the left hand side collector.

$$v_{CR} = a_{DM}(v_+ - v_-) - \tfrac{1}{2}a_{CM}(v_+ + v_-)$$

$$v_{CL} = a_{DM}(v_- - v_+) - \tfrac{1}{2}a_{CM}(v_+ + v_-)$$

If a differential output, $v_{CR} - v_{CL}$, is taken the common mode rejection is complete.

$$v_{DO} = 2a_{DM}(v_+ - v_-) \qquad (2.57)$$

2.3.8 Bipolar Transistor Amplifiers IV - Common Base

This is another connection you may come across. I shall only mention it briefly here, as it is not very common.

The biasing (not shown) must be a negative voltage, to ensure that the transistor is on. The small signal properties are best derived from the 'α/g_m' model.

$$a_v = - g_m v_{BE} R_L / v_{BE} = - g_m R_L \qquad (2.58)$$

$$r_{in} = \alpha/g_m \approx 1/g_m \qquad (2.59)$$

$$r_{out} = R_L \qquad (2.60)$$

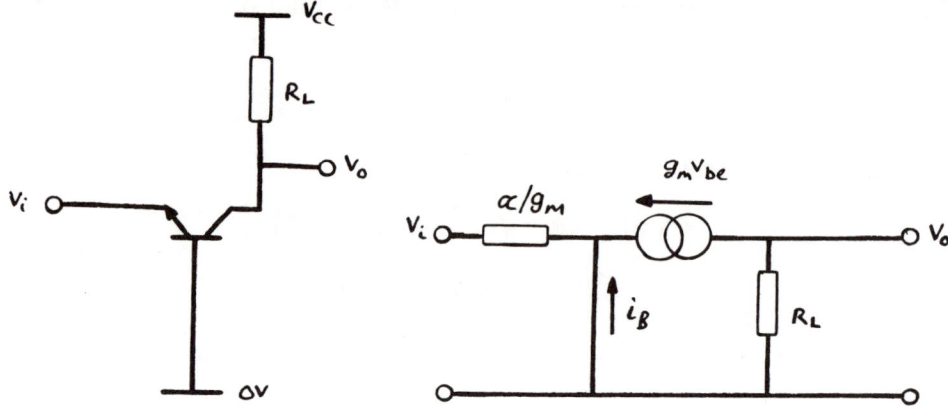

Figure 2.38 Common base amplifier and small signal model

2.3.9 Current Sources and Current Mirrors

The properties of transistors which enable them to be used as current sources include the fact that over a very large range of collector currents $V_{BE} \approx 0.7$ V and that the collector current is independent of V_{CE}.

An ideal current source causes a fixed current to flow through a load of any magnitude. Such a situation can be made by an active circuit which senses the load current, and adjusts its output p.d. in order to keep the current constant. This type of current source will be discussed in a later chapter. Simpler designs can be made by exploiting the above properties.

Some Simple Current Sources

$$I_E = V_E/R_E \quad \Rightarrow \quad I = (V_Z - V_{BE})/R_E \qquad (2.61)$$

V_{BE} is approximately constant, at 0.7 V, over a wide range of currents.

Deficiencies with these Sources

The *Early* effect. It is found that, at constant I_C, there is some slight dependence of V_{BE} on V_{CE}. Thus, the current will depend slightly on the load. ($\delta V_{BE} \approx -\delta V_{CE}/1000$)

β depends on V_{CE}.

V_{BE} depends on temperature (approx. - 2 mV K^{-1}).

β depends on temperature.

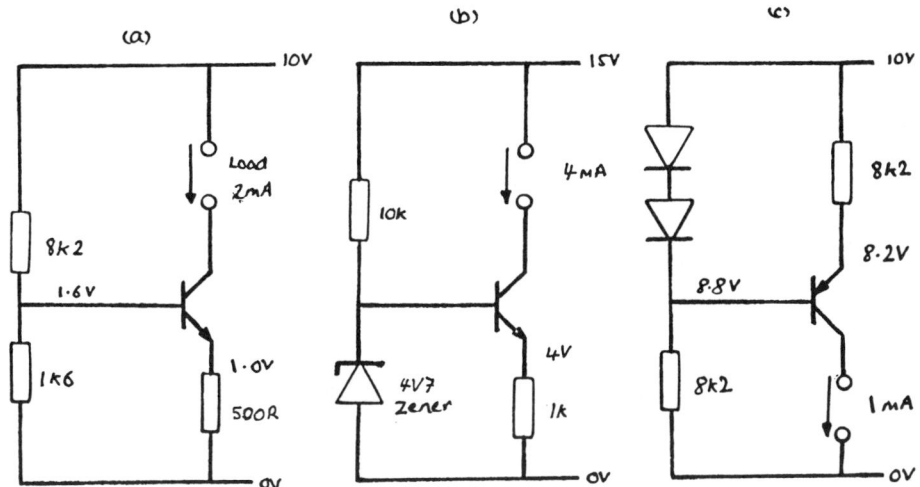

Figure 2.39 Some simple current sources

Compensated Current Sources

This circuit compensates for temperature changes in V_{BE}.

If the ambient temperature changes by +1°C, then the base-emitter drops change by -2 mV. Ordinarily this would increase the load current, because V_{BE} increases by 2 mV. However, here the emitter follower's (Q_1) base-emitter drop also changes by -2 mV, causing the base of Q_2 to drop 2 mV. Therefore the change, is eliminated.

The second circuit compensates for variations in V_{CE} of the current determining transistor, by fixing V_{CE} using a potential divider and diode-drops. This is called a *Cascode* connection. The precision resistors must be used to guarantee good performance, and for all precision circuits it is best to use high quality components with *trimmers* (small pots adjusted by a screwdriver) used somewhere to allow adjustment to compensate for errors.

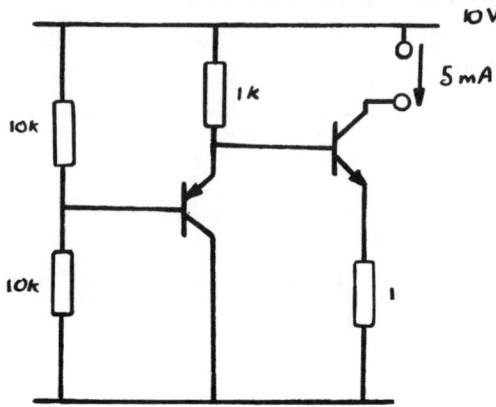

Figure 2.40 A compensated current source

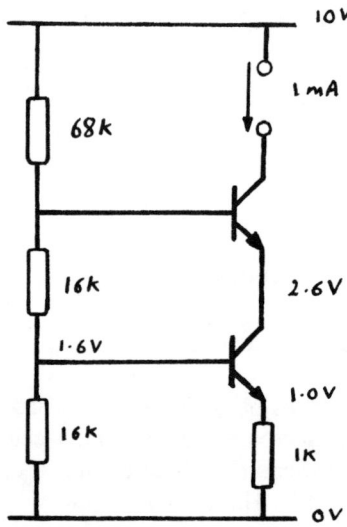

Figure 2.41 Cascode compensated current source

To avoid the effect of temperature on β, use high β transistors so that the base current contribution to the emitter current is relatively small.

Current Mirrors

The following circuit is a simple *current mirror*.

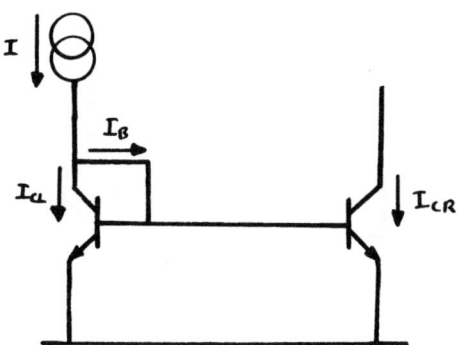

Figure 2.42 Simple current mirror

The input voltage is approx 0.6 V over a wide current range. The input current is I, of which some forms the collector current of the left hand side transistor, and for ideally matched devices the remainder splits equally to form the base currents of the left and right hand transistors. The absolute magnitude of β for the transistors does not matter, just the match of β's which must also be good.

$$I_{CL} = I - I_B$$

also (ideal match):

$$I_{CL} = \beta I_{BL} = \tfrac{1}{2}\beta I_B$$

$\Rightarrow$
$$I_B = I/(1 + \tfrac{1}{2}\beta) \tag{2.62}$$

now
$$I_{CR} = \tfrac{1}{2}\beta I_B$$

$\therefore$ from (2.62)
$$I_{CR} = \tfrac{1}{2}\beta I/(1 + \tfrac{1}{2}\beta) \tag{2.63}$$

Now, if we use transistors where $\tfrac{1}{2}\beta \gg 1$ (say 200 or more) this can be simplified.

$$I_{CR} \approx I \tag{2.64}$$

The current drawn by the RHS is that fed in on the LHS, but the RHS is isolated (to a large degree) from the LHS!

One obvious use is in the collector circuits of a long tailed pair, to force the quiescent collector currents to be equal and, therefore, match the gains of the transistors. However, the resistance of a current source is very high and you will get very high gains with such a circuit. In a later section you will see how to deal with 'near-infinite' gains to produce workable circuits.

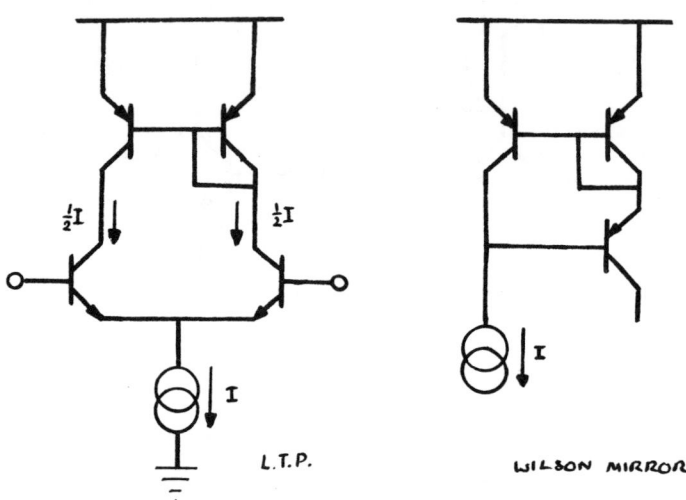

Figure 2.43 Current mirror load LTP and Wilson mirror

Before discussing a very useful aspect of the current mirror, here is a circuit of a very stable mirror, called the Wilson Mirror. The job of the extra transistor is to avoid the Early effect by fixing V_{CE} for both current determining transistors.

Multiple Output, Current Multiplying and Dividing Mirrors

If another transistor is added to the RHS of the mirror it also ends up with a collector current equal to the input current, but the base current is split between three bases, so we must change (2.63) to read:

$$I_{CR} = \beta I/n(1 + \beta/n) \tag{2.65}$$

for the general condition when we have n - 1 outputs. (i.e. single output n = 2; double output, as above, n = 3 etc.)

If, instead of taking separate outputs, the output collectors are connected together the output current is $I_{CR} + I_{CR} \approx 2I$.

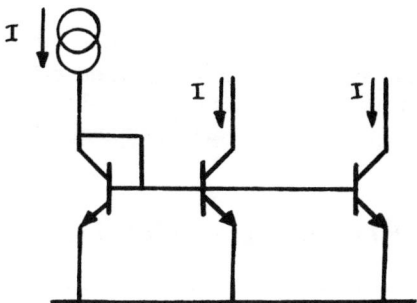

Figure 2.44 Double mirror

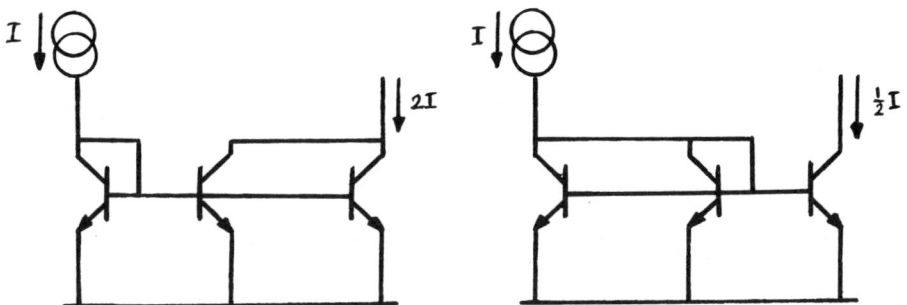

Figure 2.45 Multiplying and dividing mirrors

Alternatively, two left hand side transistors can be used, in which case:

$$I_{CR} \approx \tfrac{1}{2}I.$$

These circuits can be used to produce current sources with any multiple of the *programming current* I required (provided that $\beta/n \gg 1$). Resistors take up a lot of space in integrated circuits, whereas transistors can be made very small, and where space is at a premium, or the device must work over a large range of supply voltages, designers often replace all current sources in the IC with multiplying/dividing mirrors referenced to an external current source.

2.4 The Manufacture of Semiconductor Devices

The purpose of this section is to discuss, briefly, how semiconductor devices are formed. The main reason for this is that the operation of field effect transistors is easily explained with a knowledge of their structures. I also thought it would be useful to know how modern diodes, bipolar transistors, and integrated circuits, are made.

2.4.1 The Planar Process I - The Diode

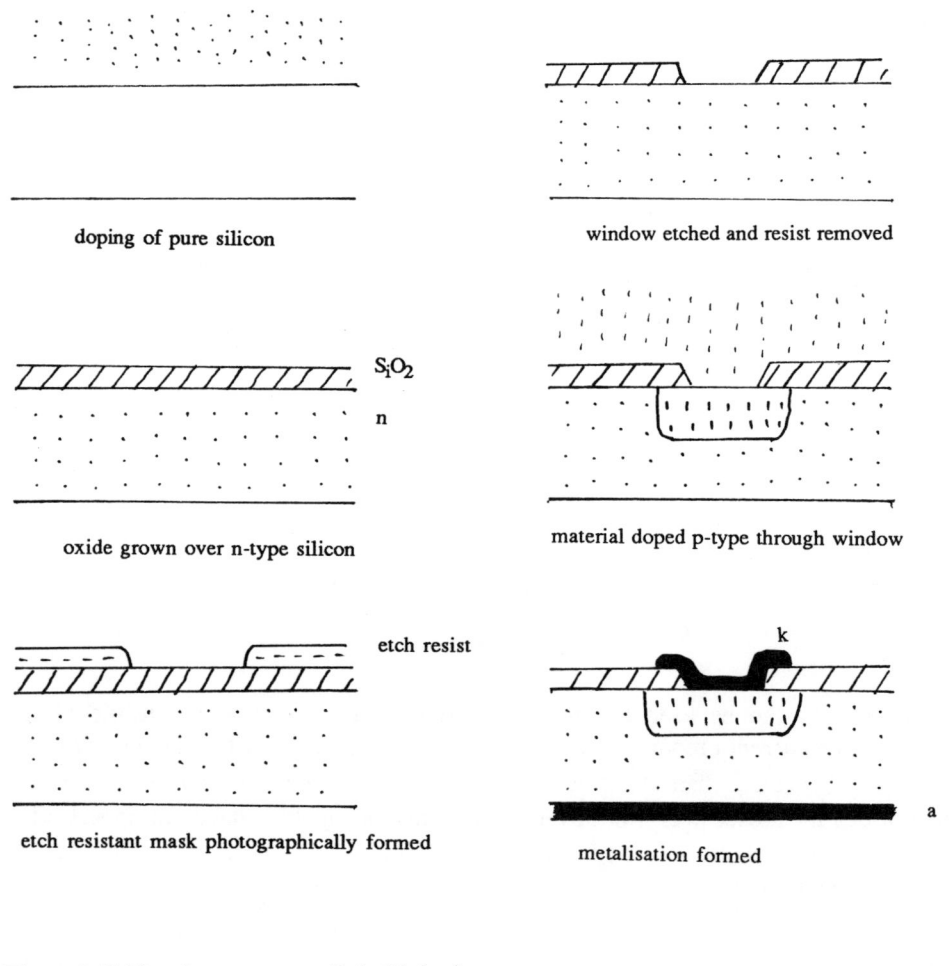

doping of pure silicon

window etched and resist removed

oxide grown over n-type silicon

material doped p-type through window

etch resistant mask photographically formed

metalisation formed

Figure 2.46 The planar process - diode fabrication

The process used is called the *Planar Process* because devices are made from a slice of pure silicon.

The pure silicon is exposed to a vapour of a donor impurity to turn the entire slice into n-type silicon. The impurity slowly diffuses (spreads) into the silicon.

Then a thin film of silicon dioxide (silica) is formed over the top. This is etched to produce a window. Then the slice is exposed to a vapour of an acceptor impurity, to form a p-type region. This junction is not abrupt, but the equations derived for an abrupt junction are fairly good models. Aluminium is deposited in the window, to form a contact.

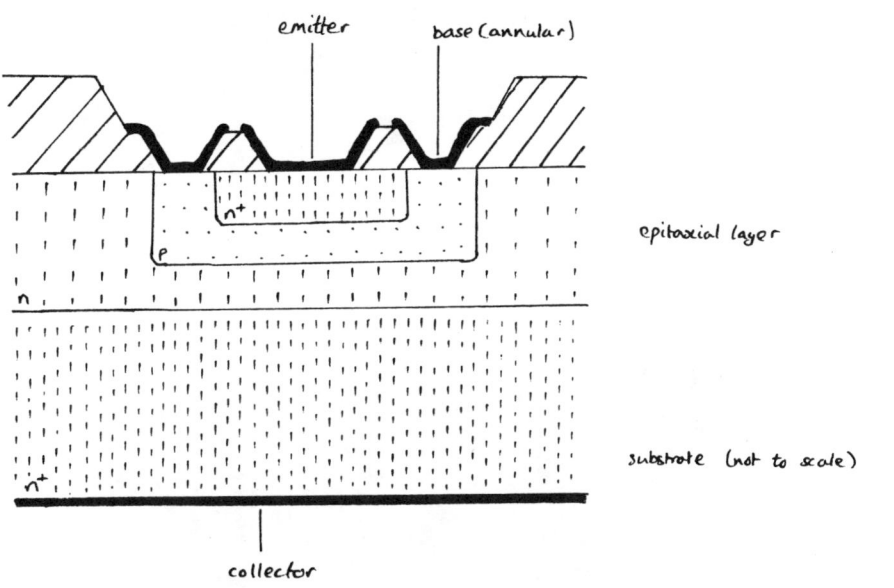

Figure 2.47 The Planar process - bipolar transistor fabrication

2.4.2 The Planar Process II - The Bipolar Transistor

First a slice of n⁺ type silicon is formed. This is called the *substrate* and forms the base for the device (but is not the *base* of the device). A thin film of n type silicon is grown on the top. This is typically 10 μm thick and is called the *epitaxial layer*. A skin of oxide is formed. A base window is etched. The base is diffused into the epitaxial layer. Oxide is regrown, followed by etching to form the emitter window. The emitter is diffused into the epitaxial layer. Windows are formed, and aluminium deposited, to make connections.

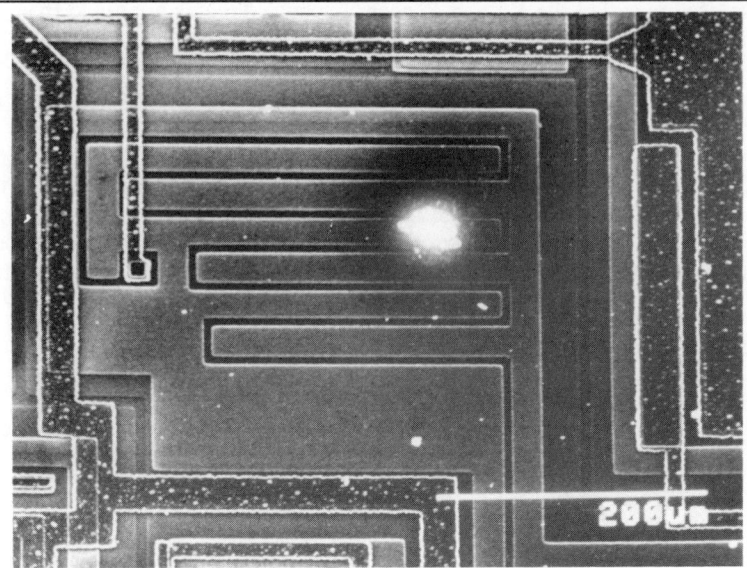

Figure 2.48 S.E.M. photograph of an integrated resistor. The 'blob' is a spec of dirt charged up by the scanning electron beam.

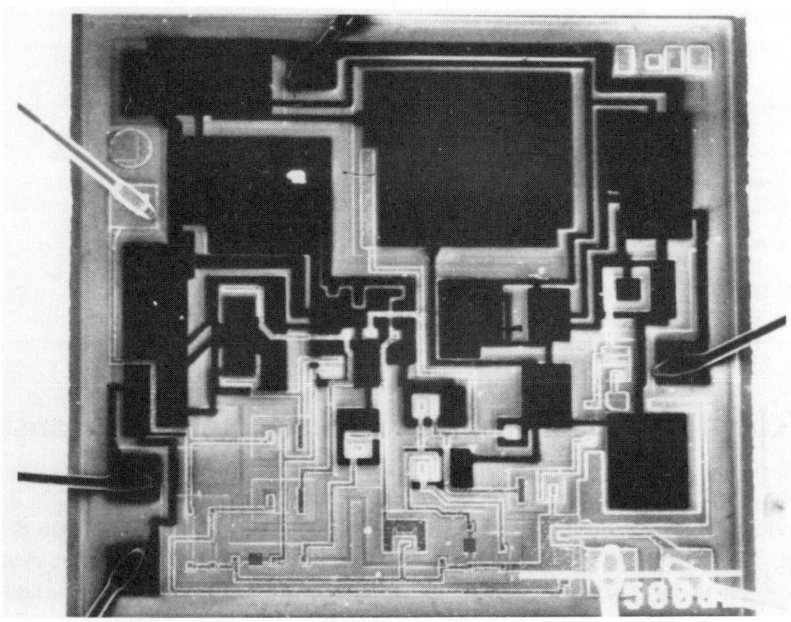

Figure 2.49 S.E.M. photograph of 741 op. amp. The S.E.M is used in 'voltage contrast mode' - positive areas are black and negative are white. The large black square in the top centre is a capacitor. The positive and negative contacts are visible.

2.4.3 Formation of Resistors and Capacitors

Resistors are made from thin films of silicon in the epitaxial layer. The actual length and thickness of the film determining the resistance. Resistors can take up a lot of space.

Capacitors are generally formed from reverse biased diodes (as in varactor diodes), or from a metal-oxide-silicon structure (MOS).

However, capacitors over ≈ 50 pF require large areas of silicon, and so the designer will take steps to eliminate the number of large value capacitances in any integrated circuit

2.5 Field Effect Transistors

2.5.1 Junction Gate Field Effect Transistors

The JFET (or JUGFET, or sometimes just FET) is a very simple device whose operation is best explained by examining its structure.

The device in figure 2.50, formed in a slice of p-type silicon, is called an *n-channel* JFET because the conduction takes place in a thin strip of n-type silicon.

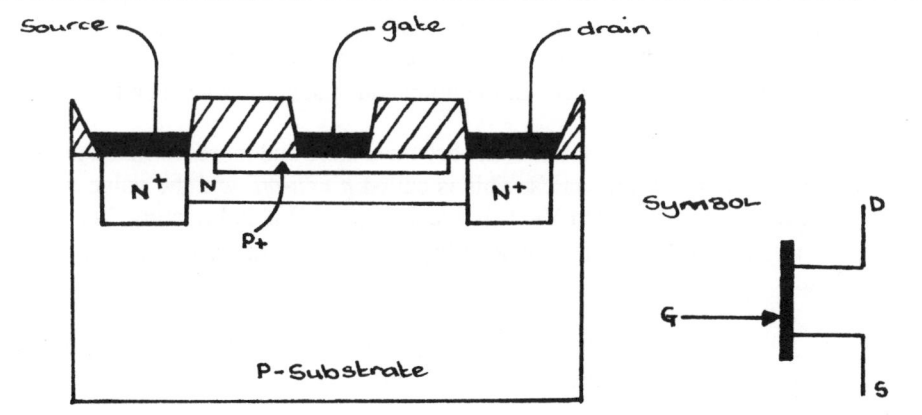

Figure 2.50 Structure of a JFET

Essentially it's a diode with a small anode and large cathode, to which two contacts are made. The anode is called the *gate*. The cathodes are called the *source* and *drain*. Generally, the drain is at a higher potential to the source, but the device is symmetrical and either cathode could be used as source/drain. Actual devices are manufactured to make the drain-gate capacitance less than the drain-source capacitance, so it is best to stick to the labels offered by the manufacturer, but you could interchange the terminals and the circuits would still work - unlike bipolar transistors in which the diodes allow conduction in one direction only.

In a circuit you always reverse bias the diode by pulling the gate-source voltage, V_{GS}, negative. This creates a depletion layer with a width dependent on V_{GS}.

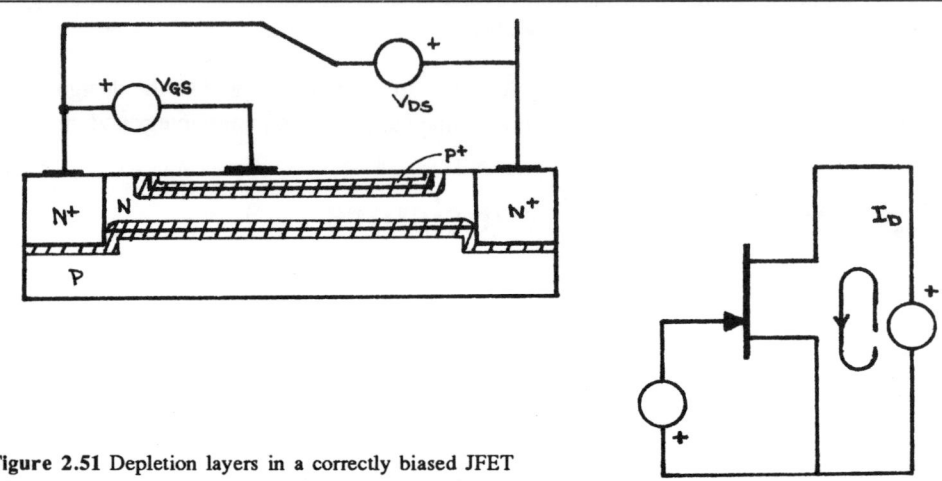

Figure 2.51 Depletion layers in a correctly biased JFET

The depletion layer is an insulator, so V_{GS} in controlling the width of the layer alters the conductivity of the *channel*, which is the strip of n-type semiconductor. The depletion layer is distorted towards the drain because, in normal circuit configuration, V_{DS} is positive, so the reverse bias is greater at the drain end than at the source end.

As V_{GS} becomes more negative the channel becomes more and more restricted until there is no channel and conduction ceases. This is called *pinch-off*, and the value of V_{GS} at which pinch-off occurs is written V_P. It is usual to quote $|V_P|$ in literature, but for an n-channel device, which is what I am discussing, pinch-off actually occurs at a negative value of V_{GS}.

The derivation of an equation describing JFET operation is harder than that for bipolar transistors, and so I am only quoting the result here.

$$I_D = I_{DSS}(1 - V_{GS}/V_P)^2 \qquad (2.66)$$

I_{DSS} is the drain-source saturation (absolute maximum) current. For JFETs it is precisely defined to be $I_{DSS} = I_D$ @ $V_{GS} = 0$. Manufacturers usually only quote $|V_P|$ but for an n-channel device V_P is a negative number and I_D looks like $(1 + x)^2$.

The good news concerning JFETs is that, because the diode is always reverse biased, only *leakage current* flows out of the base. Thus, the input impedance is very high (usually considered infinite), which makes JFETs ideal for amplifier input stages. The equivalent to the emitter follower, the source-follower, is a very useful circuit because of this.

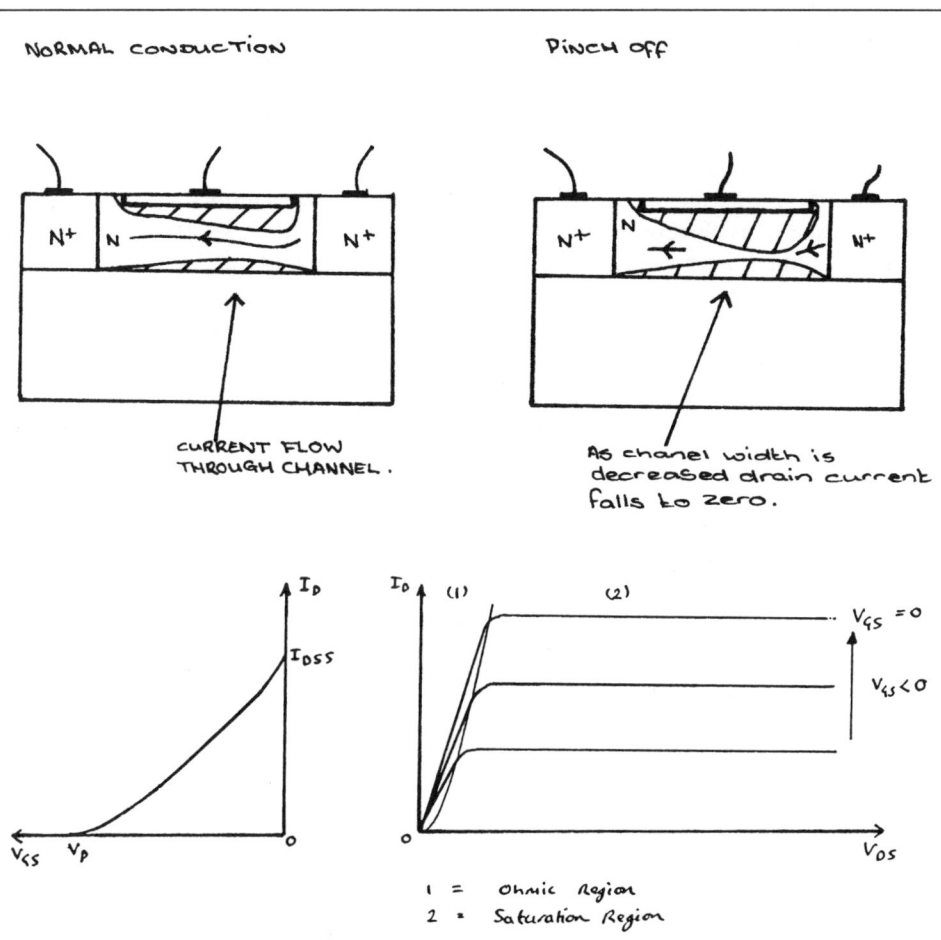

Figure 2.52 Controlling drain current using depletion layers

The bad news is that both I_{DSS} and V_P are badly defined parameters (i.e. they vary from device to device as β, does for bipolars) so check your data book when designing circuits or, better still, order in bulk and use a FET tester circuit to identify the best device for any particular circuit.

In the p-channel JFETs the channel is made of p-type semiconductor and so, to reverse bias, you make V_{GS} positive, and in the circuit V_{DS} is negative. For p-channel devices change the signs of I_{DSS} and V_P.

2.5.2 Metal-Oxide-Semiconductor Field Effect Transistors (MOSFETs)

Since the junction in a JFET is not actually used, except to apply a controlling field, an actual pn-junction is not required. In the n-channel enhancement mode MOSFET (or IGFET for Insulated Gate FET) an insulating strip of SiO_2 is formed between the metal gate and the p-type substrate. Hence the name "MOS".

When a positive potential is applied to the gate, a charge separation occurs in the *dielectric* (the insulating silicon dioxide) and its *underside* acquires a positive surface charge. This attracts electrons out of the substrate area and, when a sufficiently high gate voltage is applied, an n-type channel is formed just below the dielectric. This allows conduction to occur in the device. Because increasing the gate-source voltage increases channel conductivity, the device is called an enhancement mode transistor.

The device has the symbol given in figure 2.56. The fourth terminal is the body, or substrate connection, and must be connected to some potential which ensures the correct action of the device. In almost all cases it is internally connected to the source, which ensures such an action, and so it is safe to just ignore it. The bold bar, which symbolises a channel in transistor symbols, is broken because the channel does not form, as shown above, unless the device is active. P-channel enhancement mode MOSFETs can also be formed, and they conduct when a negative potential is applied between the gate and source.

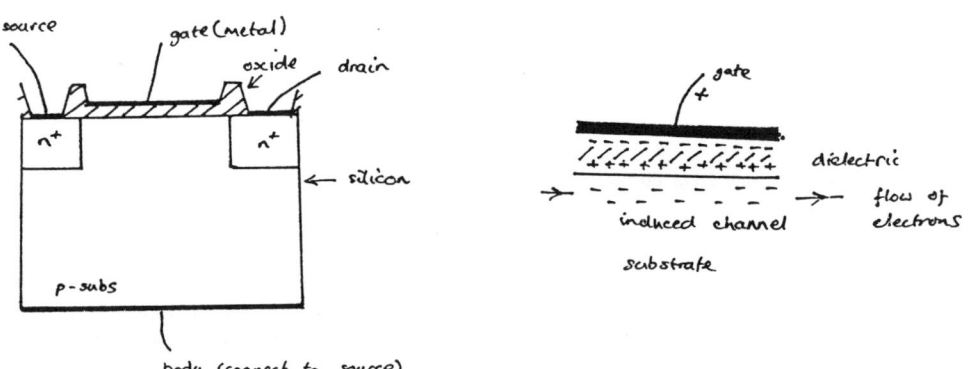

Figure 2.53 Structure of an enhancement mode MOSFET when biased correctly

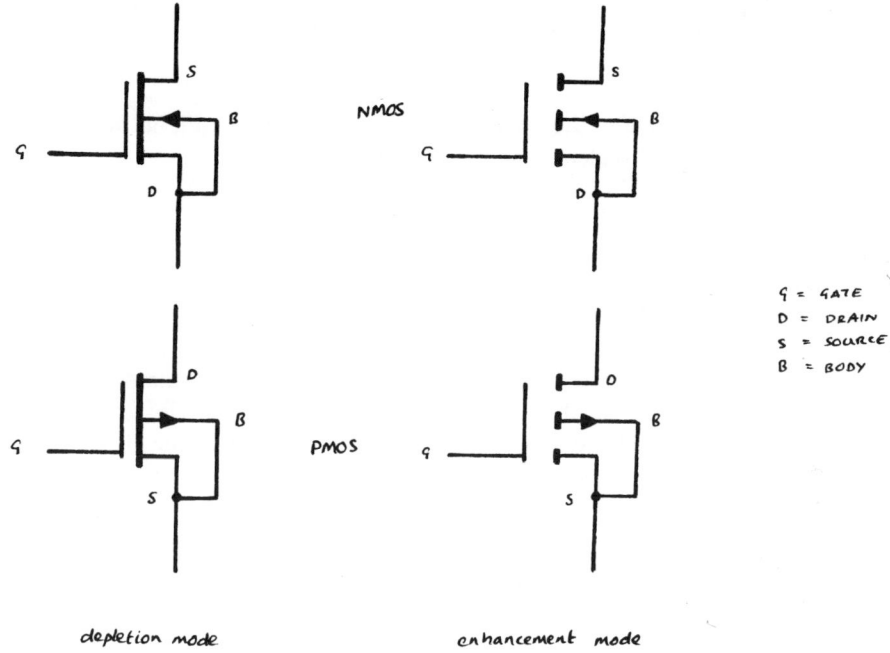

Figure 2.54 MOSFET Symbols

The forward transfer and output characteristics of p-channel and n-channel enhancement mode MOSFETs, are best analysed using two separate expressions. In the *linear* or *Ohmic* region, the equation which applies is (2.70).

$$I_D = 2M\{(V_{GS} - V_T)V_{DS} - \tfrac{1}{2}V_{DS}^2\} \qquad (2.67)$$

The onset of saturation occurs when $\partial I_D/\partial V_{DS}$ first is 0.

from (2.67) $\qquad\qquad \partial I_D/\partial V_{DS} = 2M\{(V_{GS} - V_T) - V_{DS}\}$

$\Rightarrow \qquad\qquad\qquad V_{DS} = V_{GS} - V_T \qquad\qquad (2.67)$

In the saturation region, I_D is no longer dependent on V_{DS}, and so remains at the value it takes when the saturation region is entered - which is the value when V_{DS} is given by (2.69). Substituting this result into (2.69) gives:

$$I_D = M\{2(V_{GS} - V_T)^2 - (V_{GS} - V_T)^2\}$$

$\Rightarrow \qquad\qquad\qquad I_D = M(V_{GS} - V_T)^2 \qquad\qquad (2.68)$

95

Here V_T is referred to as the *threshold voltage*, which is the voltage at which the device starts to conduct. M is a 'constant'. The magnitudes of these constants are properties of the actual device, and the signs must be adjusted so that the equation matches the forward transfer characteristic.

The absolute magnitude of the constants are not really important, but what is significant is the dependence of M on temperature.

We find: $$M \propto T^{-3/2}$$

The drain current is inversely proportional to (the cube of the root of) the temperature. The current in a bipolar transistor is proportional to (i.e. increases with) temperature. So, if the temperature increases the current will increase. This means that the power dissipated by the device will increase, which increases the temperature, etc. This behaviour can result in the destruction of the device. However, the inverse proportionality for MOSFETs (also true for JFETs) means that, if the temperature increases the current will actually decrease. Thermal runaway will not occur for FETs.

In a later section I will discuss how to use the above expressions, including what to do with the data on a manufacturer's data sheet, but before that I shall discuss the depletion-mode MOSFET.

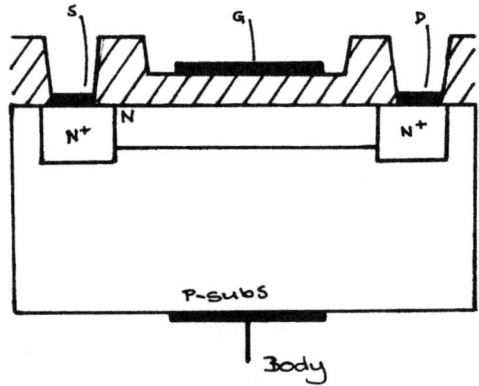

Figure 2.55 Structure of n-channel depletion mode MOSFET

With V_{DS} positive, in the n-channel device, the gate is brought to a lower potential than the source. The resulting charge separation in the dielectric causes the underside of the dielectric to acquire a negative surface charge. This repels the electrons in the channel and decreases, or depletes, its conductivity.

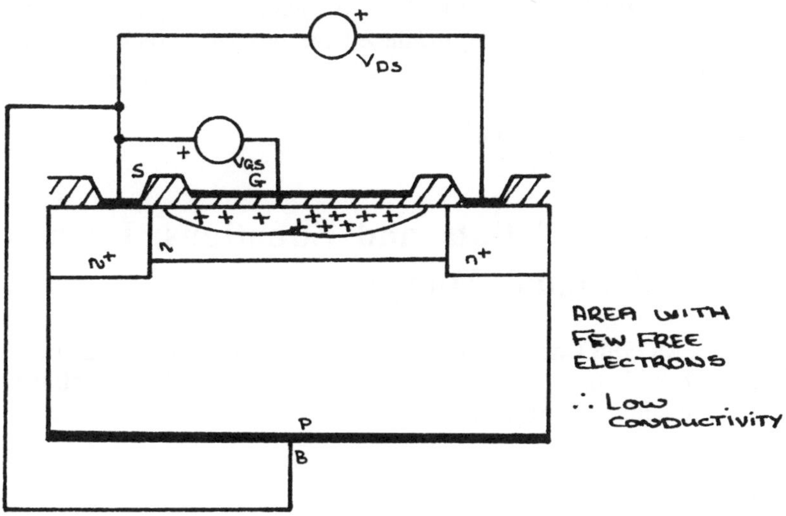

+
V_{DS}
+
V_{GS}
S
G
n+
n
n+
AREA WITH FEW FREE ELECTRONS
∴ LOW CONDUCTIVITY
P
B

Figure 2.56 Structure of the n-channel depletion mode MOSFET when conducting

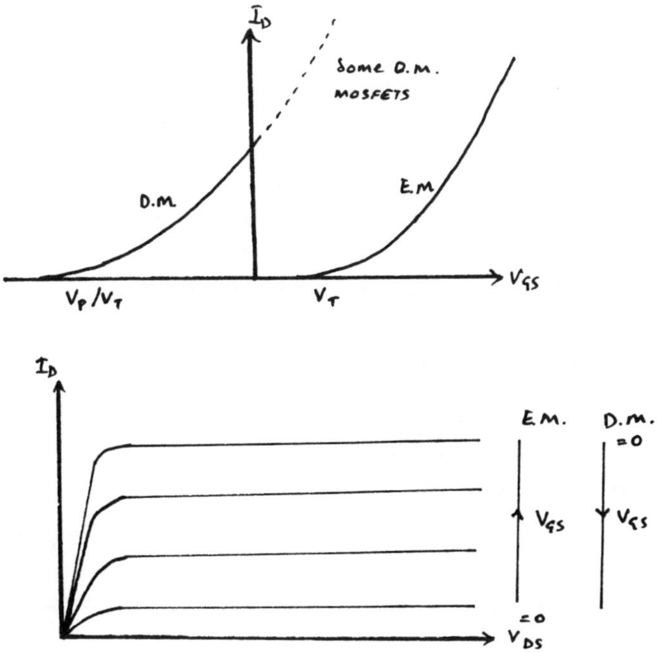

I_D
Some D.M. MOSFETS
D.M.
E.M
V_{GS}
V_P/V_T
V_T
I_D
E.M.
D.M.
=0
V_{GS}
V_{GS}
=0
V_{DS}

Figure 2.57 Characteristics of depletion and depletion-enhancement mode MOSFETs

The channel bar in the symbol is not broken, because the channel is always present. The device has the characteristics marked (a) in figure 2.57. Some devices can also be operated in the enhancement mode, and these have the characteristics marked (b) in figure 2.57. Depletion mode MOSFETs are described by the same equations as enhancement mode devices, but with the signs of the constants changed to give conduction in the correct quadrant of the V_{GS}-I_D graph.

2.5.3 Using FET Data and Equations, Including Small Signal Models

When using a MOSFET as an amplifier it will normally be in the saturation region, so equation (2.69) will apply. For a depletion-mode MOSFET, I_{DSS} is defined as the drain current (in saturation region) when $V_{GS} = 0$.

From (2.69)
$$I_{DSS} = MV_T^2 \tag{2.70}$$

$\Rightarrow$
$$I_D = I_{DSS}(1 - V_{GS}/V_T)^2 \tag{2.71}$$

This is of identical form to the JFET equation (2.66). In fact, JFETs are also referred to as depletion mode FETs because of this similarity. I_{DSS} is hard to define for an enhancement mode device so, instead, it is usual to define $I_{D(ON)}$ which is the drain current at some large value of V_{GS}, $V_{GS(ON)}$.

Thus, for enhancement mode MOSFETs:

$$M = I_{D(ON)}/(V_{GS(ON)} - V_T)^2 \tag{2.72}$$

The reason for all this is to avoid the use of M, which is not quoted by manufacturers.

Data Quoted in Manufacturers Data Sheets

For a JFET or d.m. MOSFET, the manufacturer will state:

BV_{GSS}	Gate-Source breakdown voltage (voltage at which PN-junction at gate breaks down)	typ. 20-60 V
I_{DSS}	appears in eqn (2.51.1)	few mA
V_P	pinchoff voltage = $V_{GS(OFF)}$	few V

(also device capacitances, maximum power dissipation, etc.)

For an enhancement-mode MOSFET:

$R_{DS(ON)}$	resistance of channel when $I_{DS} = I_{DS(ON)}$ (discussed later)	few 100 Ω
V_T	threshold voltage = $V_{GS(th)}$	few V
$I_{DS(ON)}$	as in equation (2.53.3)	few mA
BV_{DS}	Drain-Source breakdown voltage	10-40 V
BV_{GS}	Gate-Source breakdown voltage	10-40 V

Small Signal Models

Just like for the bipolar transistor, we find that the forward transfer characteristic of a FET is not linear, but for a signal which is a small signal superposed on a large (biasing) signal it is approximately linear. The small signal mutual conductance, g_m, is used for a FET just as it is for a bipolar transistor.

For JFETs:
$$I_D = I_{DSS}(1 - V_{GS}/V_P)^2$$

and so:
$$g_m = \partial I_D/\partial V_{GS} \approx i_D/v_{GS}$$

I_{DSS} and V_P are both constants which can be combined into the constant k.

$$I_D = k(V_{GS} - V_P)^2 \tag{2.73}$$

$\Rightarrow$
$$\partial I_D/\partial V_{GS} = 2k(V_{GS} - V_P) = 2\sqrt{(kI_D)}$$

$\Rightarrow$
$$g_{mJFET} = 2\sqrt{(kI_D)} \tag{2.74}$$

Because of the similarity between the JFET and depletion mode MOSFET equations, and the similarity between enhancement and depletion mode MOSFETs, it can be seen immediately that:

$$g_{mMOSFET} = 2\sqrt{(MI_D)} \tag{2.75}$$

k and M are constants for any particular device (at constant temperature), but because of the variation of FET parameters they will vary from one device to another. For this reason, it is usual for manufacturers to quote g_m at some specific value of I_D. If this is the case, then g_m at any other value of I_D is given by (2.75).

$$g_m = g_{m0}\sqrt{(I_D/I_{D0})} \tag{2.76}$$

For depletion mode FETs g_m is always quoted at I_{DSS}.

In normal operation i_D does not depend on v_{DS}, and so the devices appear to have a current source between drain and source. Also, I_G (gate current) is insignificant, and so the gate-source and gate-drain resistances are assumed to be infinite. Sometimes g_m is written g_{fs} or y_{fs} for FETs.

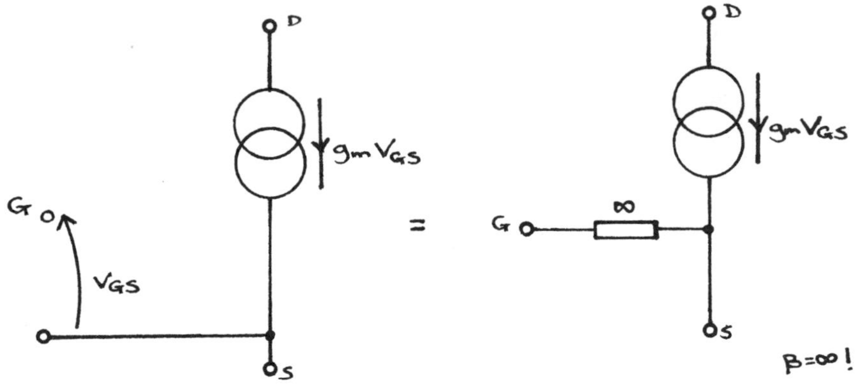

Figure 2.58 Small signal models of FETs

This is the same model as can be used for the bipolar transistor, but for FETs we have an infinite current gain, and infinite input resistance. The correspondence of models would suggest that common drain (source follower), common source and common gate amplifiers can be built with FETs in the same way that common collector (emitter follower), common emitter and common base amplifiers can be built with bipolar transistors.

Although the small signal models are the same, there are differences between FET and bipolar transistor amplifiers. Typical transconductances for FETs are millisiemens, mS, (the siemens is the unit of conductance = 1/ohms, symbol $S = 1/\Omega$; in some texts the unit is the mho, which is ohm backwards!), whereas for bipolar transistors they are a few 100 mS. Bipolar amplifiers are capable of high gains at low output impedances, but FET amplifiers are not. The large signal behaviours are also different, so the distortion will differ.

2.5.4 FET Current Sources

One use of FETs is a simple, fairly stable, current source. The most basic is analysed below.

The obvious simplicity of this circuit is a major advantage. Normally JFET, or a depletion mode MOSFET, would be used. For a current sink, using a n-channel JFET:

$$I_D = I_{DSS}(1 - V_{GS}/V_P)^2$$

Now here $V_{GS} = 0 \Rightarrow$ $\qquad\qquad I = I_{DSS}$ $\qquad\qquad\qquad\qquad$ (2.77)

The drawback is the variation in I_{DSS}. However, some manufacturers do produce two lead devices to act as current sources, as above. They go by the name of *current regulator diodes*, and the manufacturer sorts large numbers to give I_{DSS} predictability. The tolerance of a typical series (1N5283 - 1N5313) is 10%.

Source Self Biased FET Current Source

To make an adjustable, fairly stable, source you must put a stable voltage across the gate-source terminals. This could be done by using a V_{BE} reference, for example, but why not put a resistor in series with the source and exploit the stability of the current to stabilise the voltage.

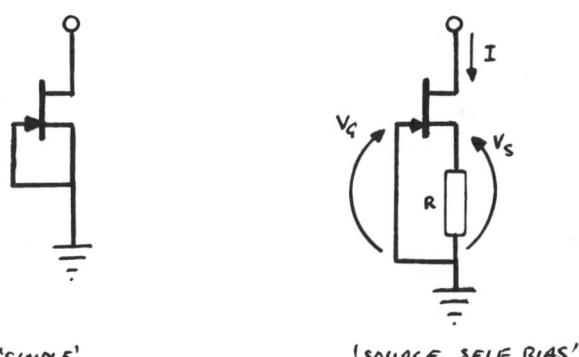

'SIMPLE'   'SOURCE SELF BIAS'

Figure 2.59 JFET current source

In the above circuit $V_{GS} = V_G - V_S = V_S - IR - V_S = - IR$

in (2.66) $\Rightarrow$ $\qquad\qquad I/I_{DSS} = (1 + IR/V_P)^2$

$\Rightarrow$ $\qquad\qquad IR/V_P = \sqrt{(I/I_{DSS})} - 1$

$\Rightarrow$ $\qquad\qquad R = V_P\{\sqrt{(I/I_{DSS})} - 1\}/I$ $\qquad\qquad\qquad$ (2.78)

2.5.5 Source Followers

This is one area in which FETs have very widespread use, and are significantly better performers than bipolar circuits in terms of input impedance. The action of the circuit is not immediately obvious.

For small signals: $v_o = g_m R_s v_{GS}$

but $v_{GS} = v_o - v_i$

$$v_o = \frac{g_m R_s}{1 + g_m R_s} v_i$$

$\Rightarrow$ (2.79)

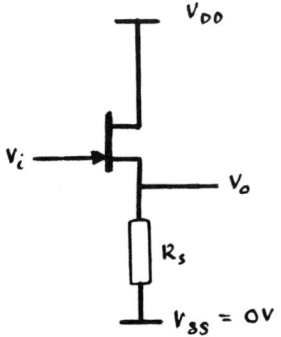

Figure 2.60 Source follower with a JFET

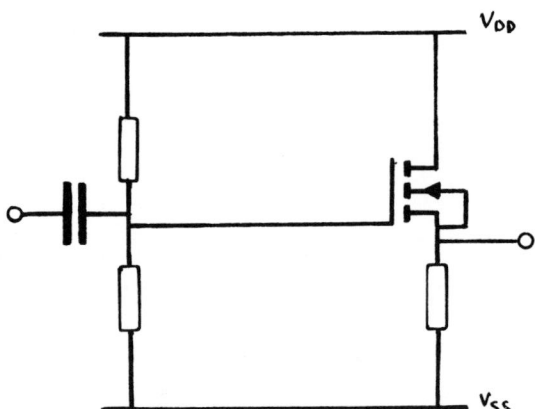

Figure 2.61 Biased MOSFET source follower

For $g_m R_S \gg 1$, $v_o \approx v_i$. This small signal amplifier must be biased. There are a number of methods, but it can be simply done with a potentiometer, just like for a bipolar transistor.

Calculation Procedure

1. Choose quiescent output voltage, usually $V_o = \frac{1}{2}V_{DD}$;

2. Find the quiescent drain current from (1), i.e. $I_D = \frac{1}{2}V_{DD}/R_S$;

3. Use (2) and equation (2.68) to find V_{GS};

4. Find $V_i = V_{GS} + V_o$.

Once you have $V_i : V_{DD} - V_i$, you can then go on to find $R_2 : R_1$ as for the emitter follower. Pick R_1 and R_2 to have this ratio using resistors of a few $M\Omega$.

Circuit Impedances

The FET will have r_{in} so high it can be assumed infinite, so the input impedance is that of the bias circuit.

$$\text{divider-biased circuit} \quad r_{in} = R_1 \| R_2 \tag{2.80}$$

The small signal equivalent circuit can be used to find r_{out}.

The transfer function is: $\quad v_o/v_i = g_m R_S/(1 + g_m R_S)$

which can be written: $\quad v_o/v_i = R_S/(1/g_m + R_S)$

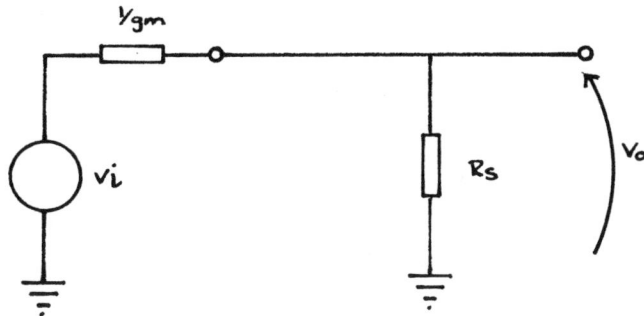

Figure 2.62 Small signal model of source follower

This is what you would get if you had the circuit of figure 2.62. The output resistor appears to be driven from a source of internal resistance $1/g_m$.

The output resistance of the circuit above is (voltage generator has r = 0)

$$r_{out} = R_S \,||\, 1/g_m$$

$$\Rightarrow \qquad r_{out} = R_S/g_m(R_S + 1/g_m) \qquad (2.81)$$

Typically, g_m is a few mS making $1/g_m$ a few hundred ohms. The source resistor will usually be a few kΩ, and so we may write:

$$r_{out} \approx R_S/g_m R_S$$

or $$r_{out} = 1/g_m \qquad (2.82)$$

Which is nice and simple, and normally not significantly different from (2.81).

2.5.6 Common Source Amplifier

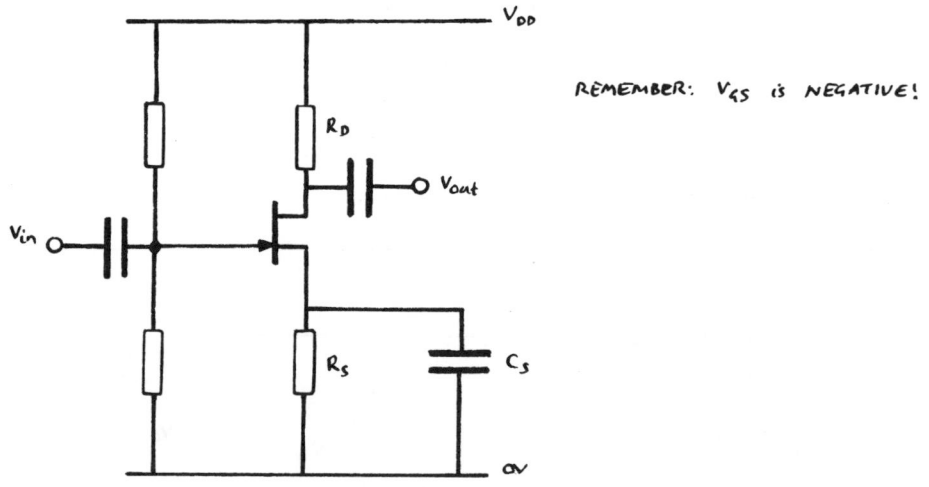

REMEMBER: V_{GS} is NEGATIVE!

Figure 2.63 Common source small signal amplifier

As I have already pointed out, circuits analogous to all bipolar amplifier circuits can be built. The circuit is a common source amplifier, analogous to the common emitter amplifier. I am using a n-channel JFET since these are generally more robust, and cheaper, than MOSFETs: but, if input impedance was critical, a MOSFET could be used.

The quiescent current is fixed to bring the device into the saturation region, and biased using a potential divider.

Small Signal Behaviour

With an input, v_G, the potential at the source is:

$$v_S = v_G - v_{GS}$$

$$i_S = v_S/R_S$$

$\therefore i_D \equiv i_S \Rightarrow$ $\qquad\qquad i_D = (v_G - v_{GS})/R_S$

and $v_{GS} = i_D/g_m \Rightarrow$ $\qquad\qquad i_D R_S = v_G - i_D/g_m$

$\Rightarrow$ $\qquad\qquad i_D(R_S + 1/g_m) = v_G$

since $v_D = - i_D R_D \Rightarrow$ $\qquad v_D = - \dfrac{R_D}{R_S + 1/g_m} v_G$

Taking $R_S \approx 0$ (i.e. proper common source amplifier) means that the small signal gain is:

$$a_V = - g_m R_D \qquad\qquad (2.83)$$

This is just as for the bipolar equivalent. However, there is a problem. Generally g_m, for FETs, is some few mS, whereas for bipolar transistors it is some 100 mS. If we want high voltage gains from a FET circuit we have to use large drain resistors, and so the circuit has a high output resistance, which is undesirable.

Due to the close correspondence between FET and bipolar transistor small signal models, other properties of the circuit can be written down without calculation.

for bipolars: $\qquad\qquad r_{in} = \beta R_E$ and $r_{out} = R_C$

$\therefore$ for FETs: $\qquad\qquad r_{in} = \infty$ and $r_{out} = R_D \qquad\qquad (2.84)$

If $R_S \gg 1/g_m$ the circuit is a common source amplifier with source degeneration, and has a gain of $- R_D/R_S$.

2.5.7 FETs as Switches and Variable Resistors

FETs as Variable Resistors

Look again at the output characteristic, in the linear region.

Provided that $|V_{DS}|$ is kept below $V_{GS} - V_T$ the curves are approximately straight lines, and the channel can be used as a resistance. A resistance whose magnitude depends on V_{GS}.

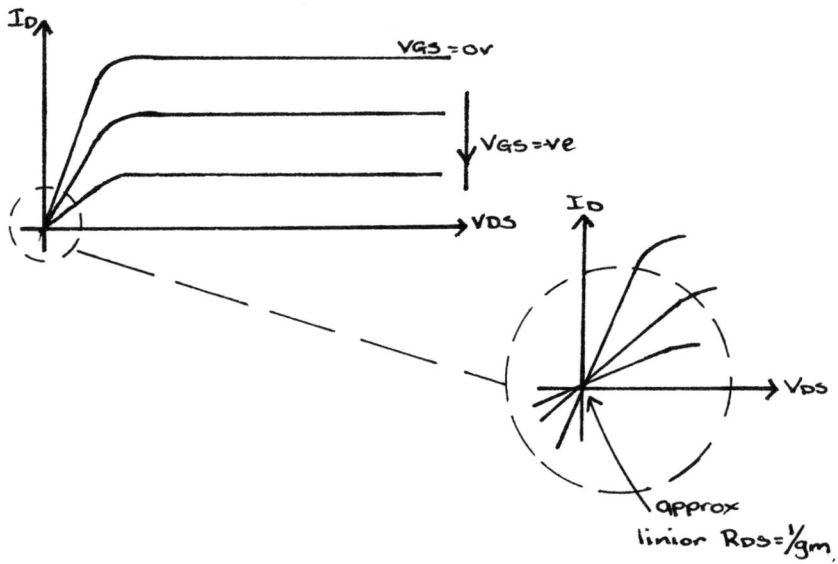

Figure 2.64 FET characteristic in the linear region

In the linear region: $I_D = M\{2(V_{GS} - V_T)V_{DS} - V_{DS}^2\}$

$\Rightarrow$ $1/R_{DS} = M\{2(V_{GS} - V_T) - V_{DS}\}$ (2.85)

For $V_{DS} \ll 2(V_{GS} - V_T)$: $R_{DS} = 1/2M(V_{GS} - V_T)$ (2.86)

Or, in terms of g_m: $R_{DS} = 1/g_m$ (2.87)

Some manufacturers quote the channel resistance at some specified voltage. The resistance at all voltages can then be deduced. e.g. if $R_{DS} = R$ at $V_{GS} = V$

$$\Rightarrow \qquad R = 1/2M(V - V_T)$$

Combining this with (2.86) enables the 1/2M to be eliminated.

$$R_{DS} = \frac{V - V_T}{V_{GS} - V_T} R$$

$\therefore$ (2.88)

There are many uses of this, such as voice controlled faders, electronic gain control (reduces RF gain of a radio, for example, if clipping distortion occurs) etc.

FETs as Switches

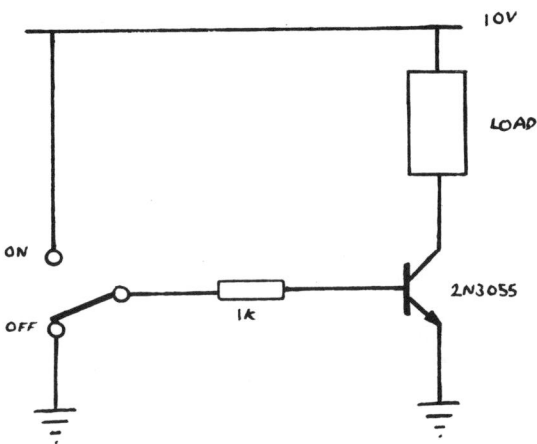

Bipolar switch

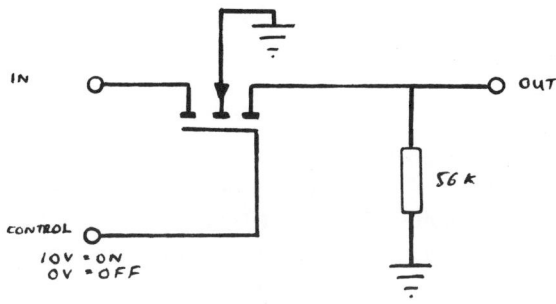

Figure 2.65   MOSFET switch

A *switch* is a device which can selectively pass currents in both directions. A bipolar transistor can be wired so that applying a potential to the base turns on, or off, a load. Since the current can only flow in one direction through the transistor, the circuit cannot be used to switch a.c. It is possible to switch a.c. using FETs, because of the symmetry between source and drain.

From (2.88) it can be seen that as V_{GS} approaches V_T (from above) then the channel resistance becomes very large. Also, if V_{GS} becomes very large then the resistance becomes very small. In reality, R_{DS} will not approach ∞, or 0, but very high, and very low, limiting resistances.

Enhancement mode FETs are particularly suitable because the channel does not form until a p.d. is applied between gate and source, and so the OFF resistance will be very high (e.g. some 10 GΩ = 10^{10} Ω). In FETs intended to be used as switches the ON resistance is typically less than 100 Ω.

The circuit of figure 2.66 is suitable. It uses an n-channel enhancement mode MOSFET and a resistor to form a pot.

When V_{GS} is brought to 0V (or lower) the fraction of the signal transmitted is approximately 1/1 000 000 of the input. When V_{GS} is brought to the supply potential the fraction transmitted is 99/100.

Care must be taken because:

The MOSFET insulation will break down when around 20V is applied across it.

In conduction the source rises to 99% of the input voltage. If the substrate were connected to the drain the device would not function properly, so it is connected to an extra lead in the transistor package. You should connect it to the lowest potential in the circuit for proper action.

The signal must not be permitted to go to the supply potential, because this would cause V_{GS} to drop towards V_T and the channel resistance would increase.

The signal cannot go to a high negative potential as this is likely to cause breakdown of gate insulation. This does not mean that current cannot flow in both directions through the switch, just that the source potential cannot be allowed to go too low.

CMOS Switches

If it is necessary to switch signals which approach the supply potentials then a circuit can be built from a p-channel and an n-channel enhancement mode MOSFET. Such a connection is called Complementary MOS, or CMOS. The only added complication is that when the control signal of the p-channel device goes to the positive supply voltage, that for the n-channel device must simultaneously go to the negative supply voltage (often earth) for proper conduction.

Such devices, called *analogue switches* or *transmission gates*, are so popular that they are manufactured in integrated circuit form, and cost about 50p for a chip containing four such gates.

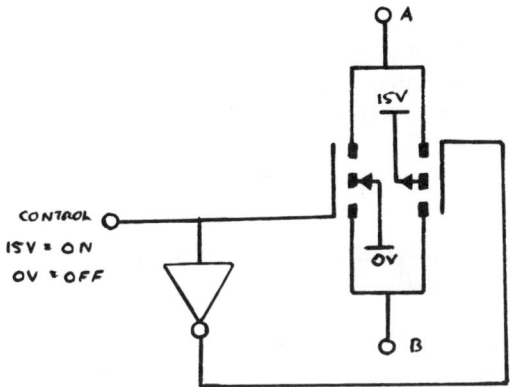

Figure 2.66 CMOS transmission gate

2.6 Unijunction Transistors

2.6.1 The Unijunction Transistor

Operation of the UJT

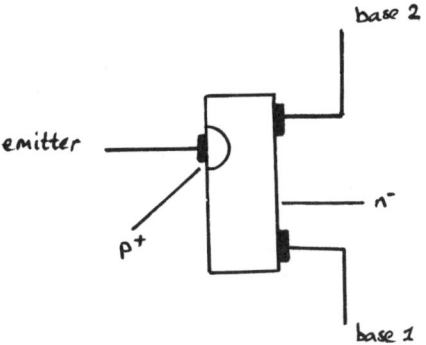

Figure 2.67 Structure of a unijunction transistor (UJT)

The unijunction transistor is a device similar in construction to a JFET. It is used mainly in making compact oscillator (pulse-generator) circuits.

The lightly doped n-type bar has a more heavily doped p-type diffused into it, as represented in the diagram. Two contacts are made to the n-type bar, called the bases. A contact is made to the p-type region, called the emitter. Unlike a JFET, the UJT is manufactured to have a definite interbase resistance, R_{BB}. This is typically 4 - 9 kΩ. The p-type region forms a rectifying junction where it is in contact with the n-type bar. When a potential, V_{BB}, is put across the bases, the n-type silicon opposite the emitter rises to a potential V_{IB}:

$$V_{IB} = \eta V_{BB}$$

η is called the *intrinsic stand-off ratio*, and depends on the position of the p-type region between the bases. It is typically 0.5.

When V_{EB1} is less than ηV_{BB} the junction is reverse biased and only leakage currents flow. When V_{EB1} is just greater than ηV_{BB} holes cross the junction (according to the diode law) and mostly combine with electrons in the base region. When V_{EB1} is greater than ηV_{BB} by about a diode drop (0.6 V) large numbers of holes are injected into the base. This potential is called the *peak-point voltage*, V_P.

$$V_P = \eta V_{BB} + V_F \qquad (2.89)$$

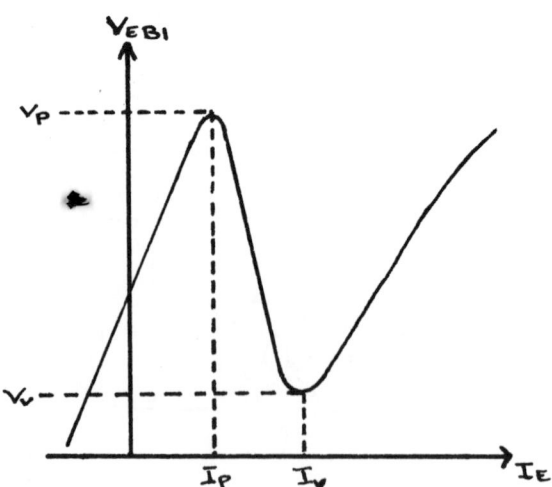

Figure 2.68 UJT characteristic

A few of the holes combine but, because the emitter is more heavily doped than the base, most are swept into the lower potential base 1 region. The effect of this is to increase the conductivity of the base 1 region, without affecting the conductivity of the base 2 region. This change in base conductivity causes V_{IB} to fall below ηV_{BB}, which increases the forward bias and increasing the effect until the resistance of the base 1 region is effectively zero. Figure 2.68 shows how the emitter current is related to V_{EB1}.

UJT Oscillators

As I have mentioned above, the main use of the UJT is in oscillator circuits. The type of oscillator made is called a *relaxation oscillator* because of its action. The obvious appeal of the circuit is the simplicity.

When power is applied the capacitor slowly charges up via the resistor. At the peak-point voltage the UJT *fires*, and starts to discharge the capacitor. As a result of UJT action the discharge is extremely rapid. The cycle then starts again.

Operating Frequency of UJT Oscillator

The potential across the capacitor is a *sawtooth* wave.

The peak emitter voltage is V_P. The minimum emitter voltage, after the initial pulse, is the minimum value of V_E when the device is conducting large emitter currents. This is the *valley point voltage*, V_V. When UJT has fired the resistance of the base 1 region falls to such a low value that the discharge time is insignificant when compared to the charge up time. The formula for operating frequency is only valid when this assumption is true.

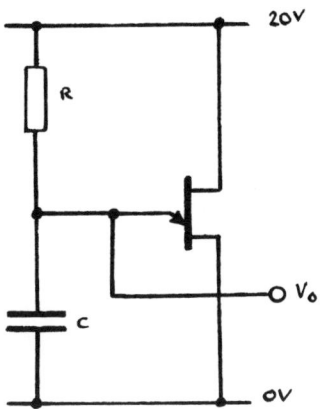

Figure 2.69 Simple UJT sawtooth oscillator

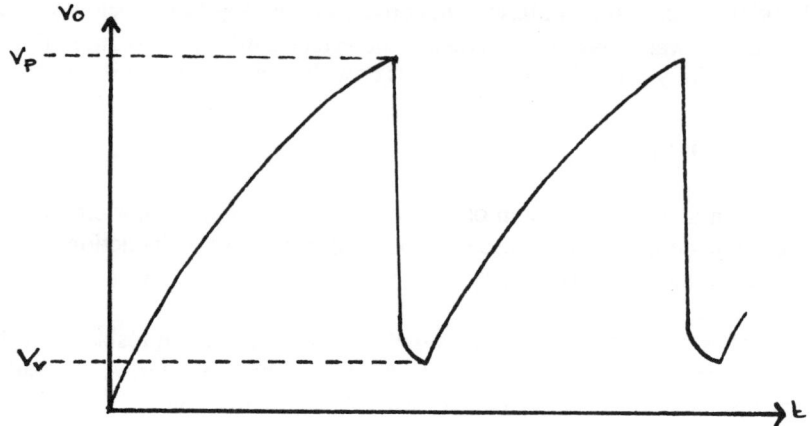

Figure 2.70 Waveform in UJT oscillator

We know that the charge-up is an exponential growth described by (2.90).

$$V = V_{BB}(1 - e^{-t/RC}) \tag{2.90}$$

This can be used to find the time taken for the capacitor to charge from V_V to V_P. Suppose that, at some time, $t = \tau$, the capacitor has V_V across it.

$$e^{-\tau/RC} = 1 - V_V/V_{BB}$$

$\Rightarrow$
$$\tau = - RC \ln (1 - V_V/V_{BB})$$

$\Rightarrow$
$$\tau = RC \ln \frac{V_{BB}}{V_{BB} - V_V}$$

At some later time, $t = \tau'$, the capacitor has V_P across it.

$\Rightarrow$
$$\tau' = RC \ln \frac{V_{BB}}{V_{BB} - V_P}$$

The period of the oscillation is $T \approx \tau' - \tau$.

$\therefore$
$$T = RC \ln \frac{V_{BB} - V_V}{V_{BB} - V_P} \tag{2.91}$$

If $V_{BB} - V_V \approx V_{BB}$, and $V_{BB} - V_F - \eta \, V_{BB} \approx V_{BB} - \eta \, V_{BB}$, then

112

$$f = -1/RC \ln (1 - \eta)$$ (2.92)

The frequency of oscillation is f. C can take any value, but R is restricted. If R is too large the emitter current will never reach I_P, and so triggering will not occur. This gives the condition:

$$R < (V_{BB} - V_P)/I_P$$

If the current flowing though the resistor is larger than a critical minimum the capacitor will be charged quicker than the UJT can discharge it, and consequently never turn off. This condition is:

$$R > (V_{BB} - V_V)/I_V$$

The timing resistor must obey the selection rule (2.95).

$$(V_{BB} - V_V)/I_V < R < (V_{BB} - V_P)/I_P$$ (2.93)

Improved UJT Oscillator Circuits

The most important improvement in the UJT oscillator is to add two small resistors in series with the bases. R_1 acts to limit the discharge current, and is usually about 50 Ω. R_1 can also be used to give an output of short *spikes* instead of the sawtooth wave taken from the capacitor.

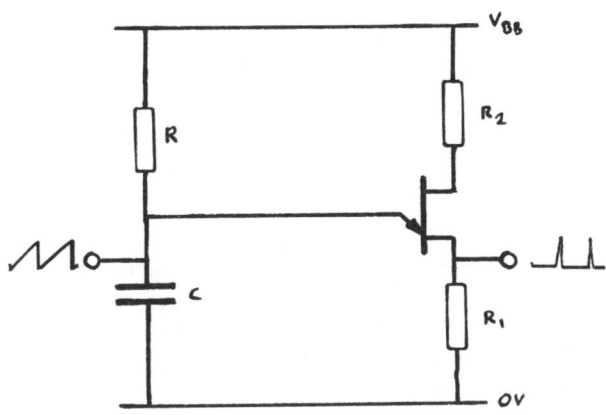

Figure 2.71 Better UJT Oscillator

113

When fired, there will be about a diode drop between the emitter and base 1 or the UJT, giving:

$$R_1 = (V_P - V_F)/I_{Emax}$$

R_2 acts to give temperature stability. For modern, annular structure, UJTs the optimum value of R_2 is:

$$R_2 = 0.015\eta \ V_{BB}R_{BB}$$

This is typically about 300Ω.

Both these resistors are going to shift V_P slightly (the exact amount is easily calculable). However, since R_{BB} is around 6 kΩ and these resistors are much less than R_{BB}, the effect is usually negligible.

Other Improvements

(a)   Replacing R with a current source for greater linearity in the output;

(b)   Adding a bipolar discharge transistor;

(c)   Using an emitter follower to reduce loading effects.

2.6.2 The Programmable Unijunction Transistor

The PUT is like a UJT in which V_P is set by an external divider. This means that the operating frequency is not dependent on variations in η.

It is a 4-layer pnpn device (a), but is best thought of in terms of the equivalent (c). The use of the equivalent is justified by (b). (d) is the symbol. The terminals are called the anode, A, cathode, K, and gate, G.

When there is about a diode drop between the anode and the gate the pnp transistor turns on. This causes a large current to flow into the base of the npn transistor, which saturates *jamming* the base of the pnp transistor to a low potential and keeping the pnp transistor on. That keeps the npn transistor on, and the device is fired. Conduction stops when the anode current drops below a critical value, which is usually about 100 µA.

A typical PUT oscillator is shown in figure 2.73. It outputs a sawtooth wave, like the UJT oscillator.

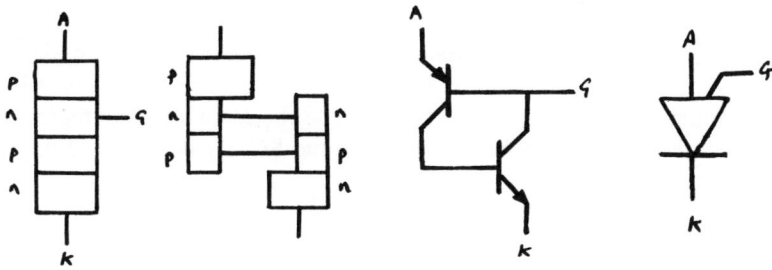

Figure 2.72 Structure of a Programmable Unijunction Transistor (PUT)

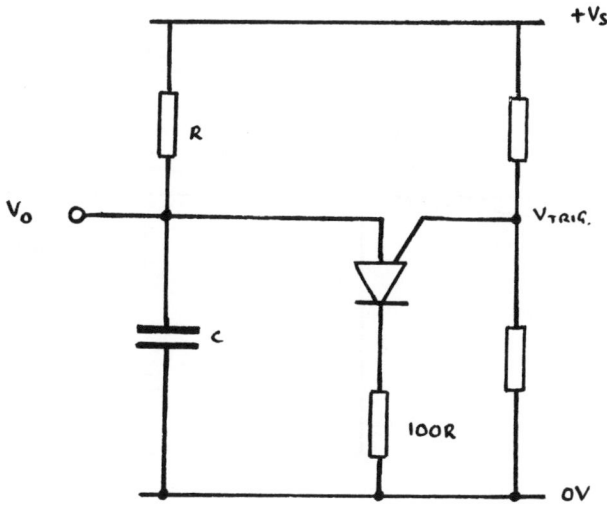

Figure 2.73 PUT oscillator

2.7 Thyristors and Power Control

2.7.1 The Reverse Blocking Thyristor Triode - S.C.R.

The silicon controlled rectifier, formally known as the reverse blocking thyristor triode, is a four layer pnpn device similar in construction to a PUT. The major differences are that the *gate* terminal of the SCR is connected to the second p-type region, whereas in the PUT it is the first n-type region, and that the SCR is designed to operate at very high powers.

When the SCR is connected such that V_{AK} is negative it has a high impedance between anode and cathode, and so passes a minimal current. This is the reverse blocking state.
If V_{AK} is made positive it stays in a high impedance, forward blocking state, until V_{AK} reaches a critical voltage called the *forward breakover voltage* V_{BO}. At this point, the SCR *fires* and the forward current increases greatly. The forward breakdown does not damage the device, but reverse breakdown does.

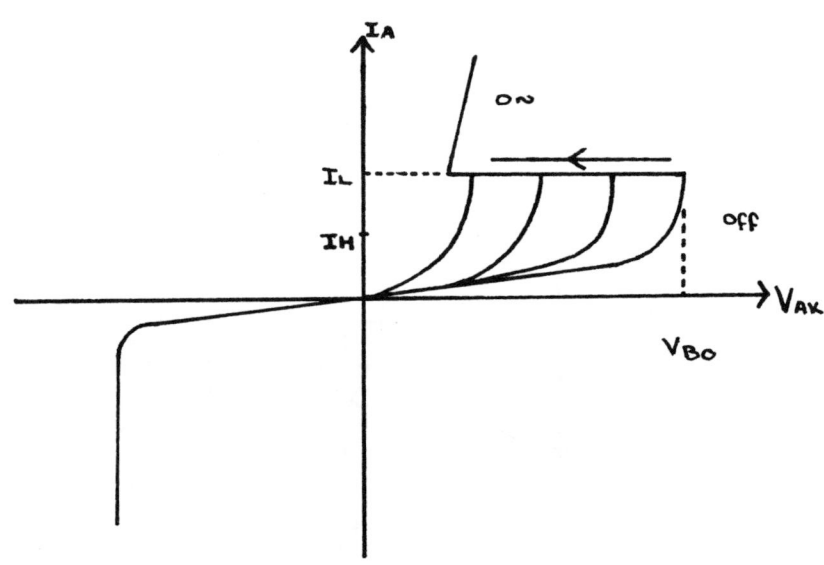

Figure 2.74 Characteristics of an S.C.R.

The SCR can be made to turn on below the forward breakover voltage by applying a trigger pulse to the gate. The magnitude of the gate current changes the voltage at which SCR firing occurs, and so can be used to *trigger* the device. This is usually done by applying a small pulse, of around 1 V, to the gate when V_{AK} is positive.

The gate pulse biases the npn transistor, formed by the lower three doped areas, into conduction. If the current drawn by the npn transistor, through the base of the pnp transistor, reaches a value called the *latching current*, I_L, the SCR turns fully on and will not turn off unless the forward current drops below the *holding current*, I_H. When the SCR is latched on the gate current can be removed and conduction will continue.

The TRIAC

The TRIAC, or bidirectional thyristor triode, is a another multilayer device. Essentially it behaves as two SCRs connected in reverse parallel with commoned gates. A positive gate pulse causes it to conduct positive current (provided that the correct potential is applied across the terminals) and a negative gate pulse causes it to conduct negative current. This device is extensively used for full wave a.c. power control because it does not have the rectifying property of the SCR.

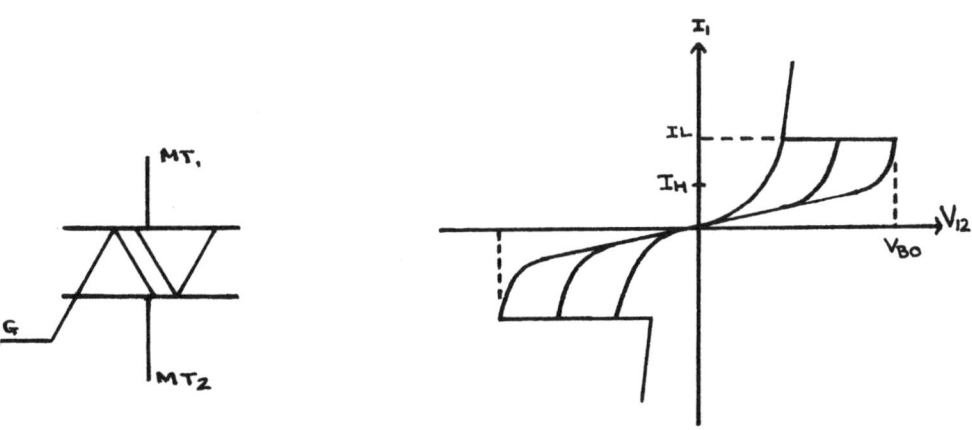

Figure 2.75 Characteristic of a TRIAC

The Gate-Turn-Off Thyristor

The GTO, or reverse blocking turnoff thyristor triode, is a slightly more sophisticated version of the SCR. The GTO will turn off when a negative pulse is applied to the gate, making it ideal for use in: d.c. switches; power inverters (turns d.c. into a.c.); and logic circuits.

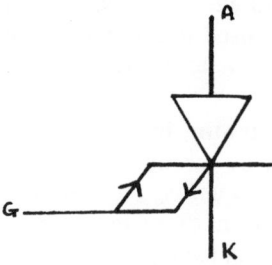

Figure 2.76 Symbol of a GTO

The Shockley Diode and the DIAC

This device is the baby brother of all thyristors. It is a pnpn structure with only anode and cathode terminals, and so its operation is that of an SCR with the gate lead cut off. It has a high impedance to forward signals until the forward breakover voltage is exceeded, when it switches to a low impedance.

The DIAC is two Shockley diodes in reverse parallel.

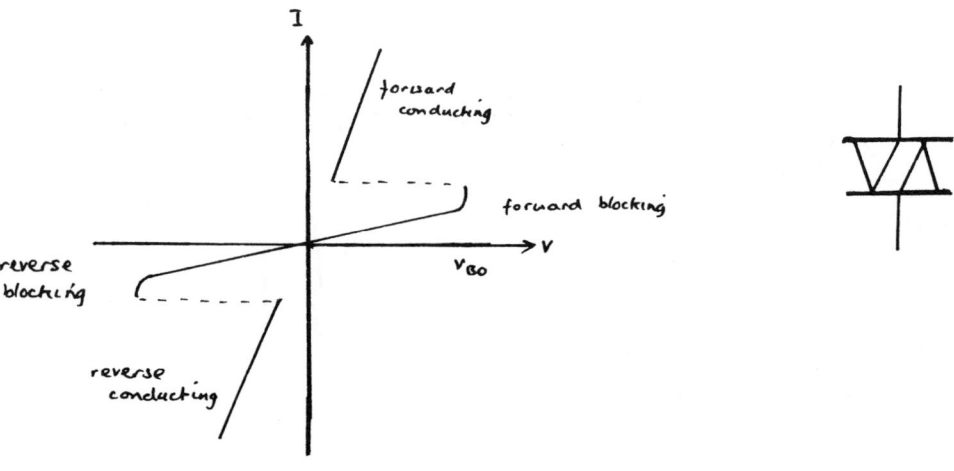

Figure 2.77 Diac Characteristic and Symbol

2.7.2 SCRs and A.C. Power Control

SCRs and TRIACs are widely used to control high power devices from a.c. power supplies. The two principal methods are *phase control* and *burst firing*.

Phase Control

For phase control a load and an SCR, or TRIAC, are connected in series to the a.c. mains. In this application the peak voltage of the supply, 360 V, must be below both the forward and reverse breakover voltages. The trigger circuit produces a pulse at the same frequency as the supply, but with a phase shift between the start of the a.c. cycle and the trigger pulse. The SCR only conducts for a proportion of the cycle and so the mean power delivered to the load is less than that if it were connected via a standard rectifier. By varying the *triggering angle* the power delivered to the load can be continuously varied.

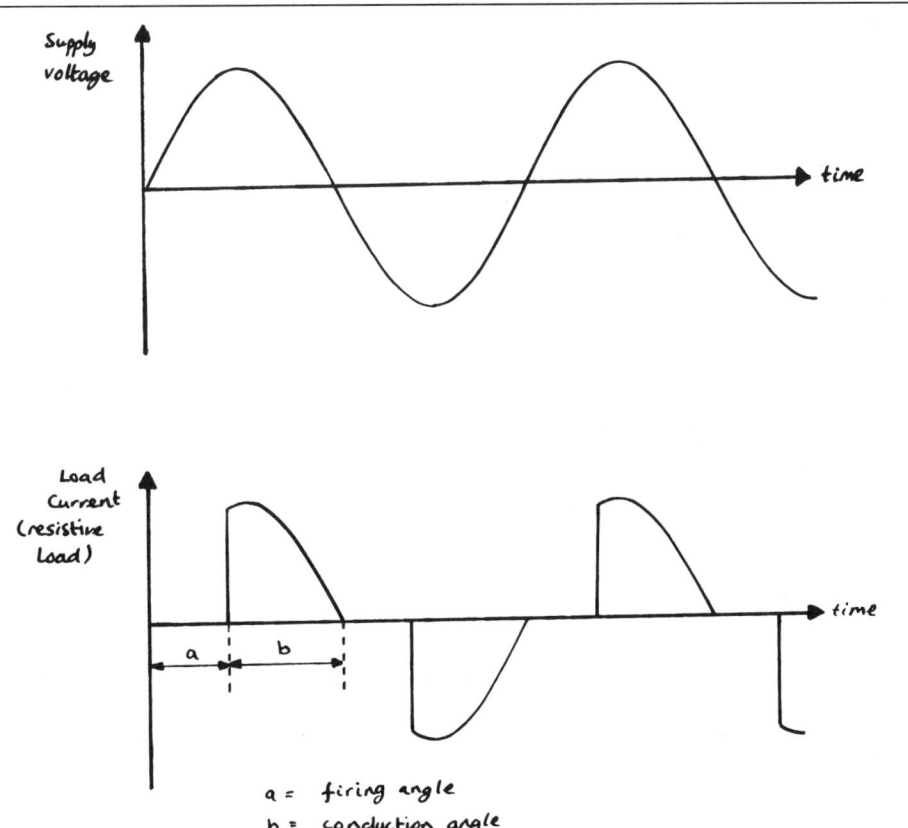

Figure 2.78 Phase control of A.C.

Burst Firing

In burst firing the principal is the same as for phase control, i.e. to vary the mean power delivered to a load by only turning it on for a fraction of the possible time, but it is accomplished differently. Instead of turning the TRIAC on for a fraction of every cycle, it is activated for n complete cycles, and then kept off for p complete cycles.

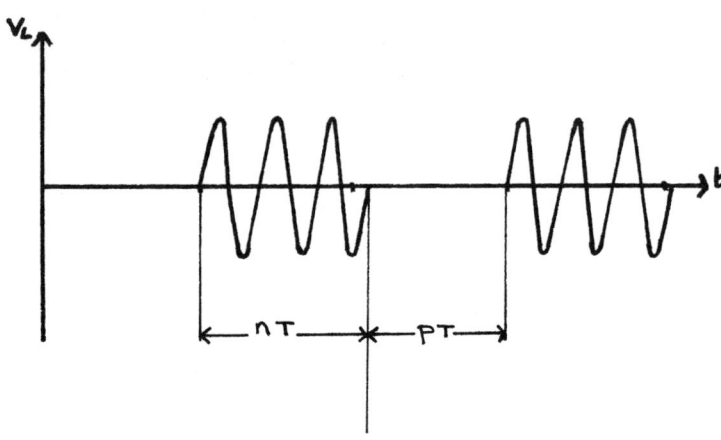

Figure 2.79 Burst firing control of a.c.

2.7.3 Controlling Inductive Loads

The behaviour of inductive loads presents a problem for the designer of power control circuits.

Turning the SCR On

When a voltage pulse is applied to an LR circuit the current does not immediately go to V/R but rises to it exponentially. There is the possibility, with a large inductive load, that the SCR anode current will not reach the latching current during the trigger pulse. This means that the SCR will never turn on.

Turning the SCR Off

Another problem occurs when trying to turn the SCR off. In a typical inductive load situation the SCR anode sits at potential, V_{AK}.

$$V_{AK} = V_S - IR - L \, dI/dt$$

When trying to turn the load off dI/dt is negative, and so V_{AK} can rise above V_S. In such a situation the SCR cannot turn off because it will continue to draw positive current. This problem is solved by the usual method of connecting a clamping diode (sometimes called a *commutating* diode) across the load. This limits the back *e.m.f.* to 0.7 V and allows the SCR to turn off.

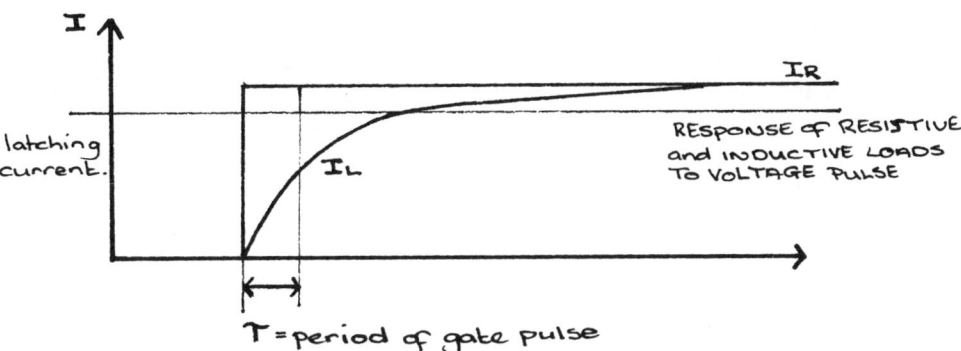

Figure 2.80 Step response of resistive and inductive loads

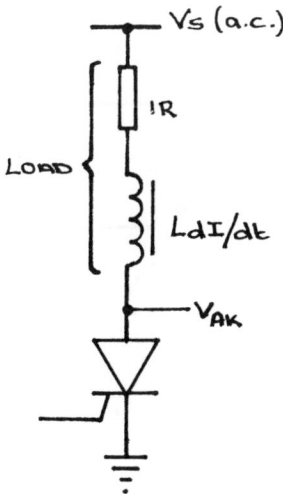

Figure 2.81 An inductive load

Questions for Chapter 2

2.1   Find the p.d. dropped by a silicon diode which is conducting:

i   10 μA
ii   100 μA
iii   1 mA
iv   10 mA

You may assume that the diode is at room temperature, which is 290 K.

2.2   Calculate the small signal resistance of the diode at those currents.

2.3   Find the change in forward voltage which is required to double the current conducted by a diode.

2.4   Using the 'α/g_m' model of a bipolar transistor calculate the small signal gain and input and output resistances of an emitter follower.

2.5   An emitter follower with a 1 kΩ collector resistor is powered from a 10 V power supply. It is stably biased so that the quiescent output voltage is 5 V. A sinusoidal input, $v_0 \sin \omega t$, is applied such that peak output is 9 V. Find v_0 and, using the method of load lines or otherwise, sketch the form of the output of the amplifier.

2.6   Design a differential amplifier to give a differential gain of 40 dB. What is the CMRR of your circuit?

2.7   Figure 2.i shows an alternative *ratio mirror* circuit. Explain its operation.

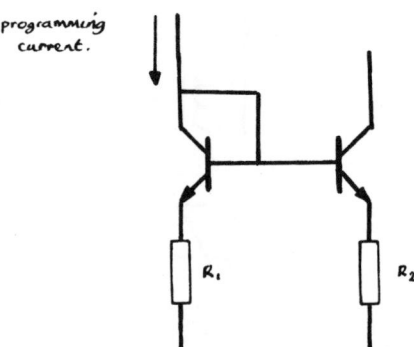

Figure   2.i

Answers to Numerical Problems

2.1   460 mV; 520 mV; 580 mV; 630 mV
2.2   2.5 kΩ; 250 Ω; 25 Ω; 2.5 Ω
2.3   17.3 mV
2.5   20 mV peak

3

OPERATIONAL AMPLIFIERS

3.1 What are Operational Amplifiers?

In the following sections I shall introduce and explain the use of operational amplifiers. Operational amplifiers are widely used and versatile devices. This is because they are made to be as close as possible to an *ideal amplifier*, and real devices differ predictably. Before I discuss the properties of the ideal amplifier I am going to introduce the concept of *feedback* which, I hope, will help explain why the particular properties of the ideal amplifier were chosen.

3.1.1 Amplifiers and Feedback

A System and an Amplifier

A *system* is any collection of devices arranged to perform a particular job. All systems contain three sub-systems:

inputs;

processing; *and*

outputs.

In analogue electronic systems (which are what we are dealing with here) the processing is done by an amplifier. An amplifier is a processor which acts on an input signal to generate an output signal.

Something happening at the input causes the generation of an input signal, s. This is modified by the amplifier to produce an output, S, which in a linear system is proportional to the input.

$$S = A_0s \tag{3.1}$$

A_0 is referred to as the *open loop gain* of the system.

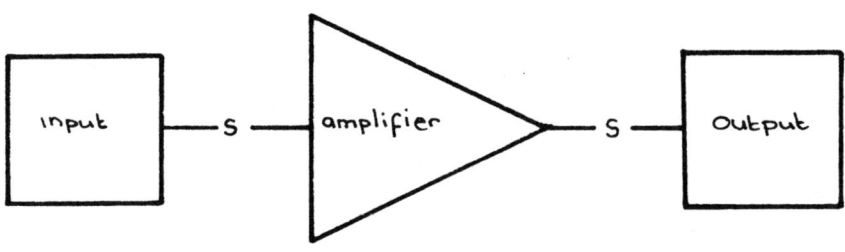

Figure 3.1 An amplifying system

A Platform Controller

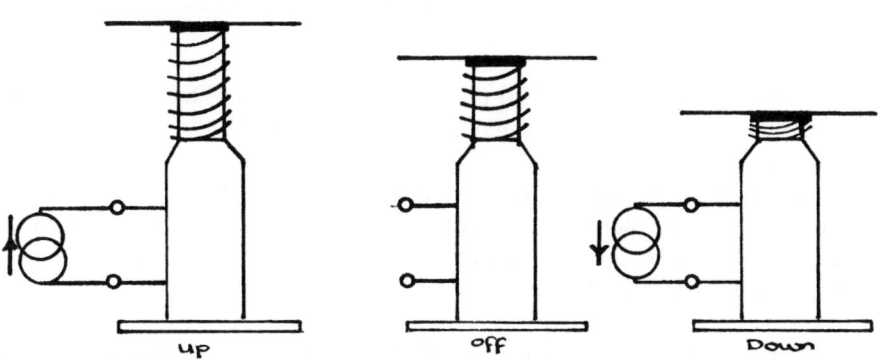

Figure 3.2 A solenoid

In order to understand this, let us consider a system in which we wish to control the position of a platform using a solenoid. A solenoid is a magnetic device in which the

125

current flowing through the solenoid controls the extension of a piston. The amount of bar sticking out of the solenoid depends on the magnitude and direction of the current flowing through the device.

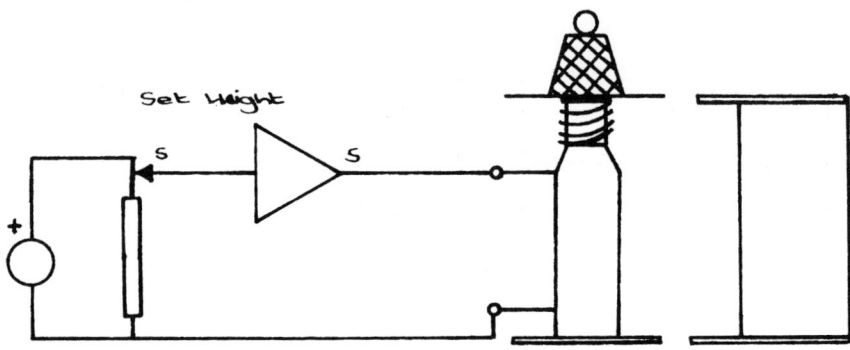

NOTE: ALTHOUGH THE SOLENOID IS A CURRENT CONTROLLED DEVICE, IT HAS AN INTERNAL RESISTANCE AND SO CAN BE OPERATED AS A VOLTAGE CONTROLLED DEVICE.

Figure 3.3 A simple platform controller

If the platform must lift heavy weights the solenoid will have to run at very high powers. To avoid the use of expensive high power control circuits I shall use a small *slider potentiometer* to control the position. A power amplifier is added so that the the voltage output, which is proportional to the slider position, can be used to control the platform position.

The gain of the system is adjusted so that when the slider is at half of its maximum position the platform is at half of its maximum position, etc.

The most significant thing that would be noticed when using the system is that, when lifting a heavy load, the platform does not rise up to the position that it would when lifting a light load. The existence of an *error*, between the desired position and the actual position, is quite a serious problem.

If the error was measured it could be combined with the input to generate a new, corrected output. This, in turn, would decrease the error and the eventual result should be that the platform is always at the correct position. Such a process is called applying *feedback* to the system.

Feedback

Negative feedback is applied when the output is sampled and compared with the desired output *and* action is taken to *decrease* the error.

Positive feedback is applied when the output is sampled and compared with the desired output *and* action is taken to *increase* the error.

The Application of Negative Feedback

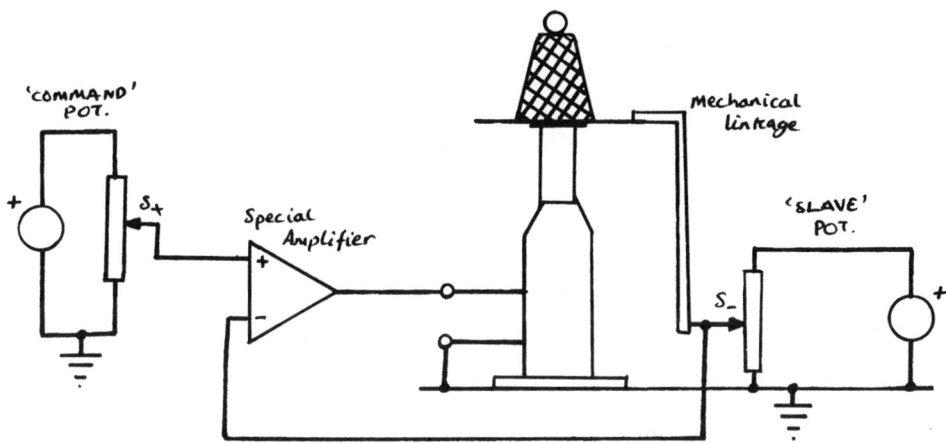

Figure 3.4 Applying negative feedback to the system

In the system studied, the output is much bigger than the input, due to the amplification, and so to compare the actual output with the desired output we must sample a fraction of the output. The system in figure 3.4 has been modified to do this. The amplifier has two terminals and its output is given by $A(s_+ - s_-)$ - it is a differential amplifier. The error signal is the difference between the input and a fraction of the output.

At a first glance you might suspect that this design would not work, because when there is no error the current to the solenoid is zero, and so the platform will drop under its own weight. However, this causes an error to occur and the circuit instantly applies a correction to compensate for it.

If the platform is too high the error is negative, and so the rod is pulled down and the platform drops; but, if the platform is too low, the error is positive, and so the platform is pushed up.

The addition of a simple negative feedback loop increases the accuracy of the system remarkably.

3.1.2 The Gain Equation and the Effects of Feedback on Circuit Impedance

The Effect of Feedback on Gain

The system with feedback is represented symbolically in fig. 3.5.

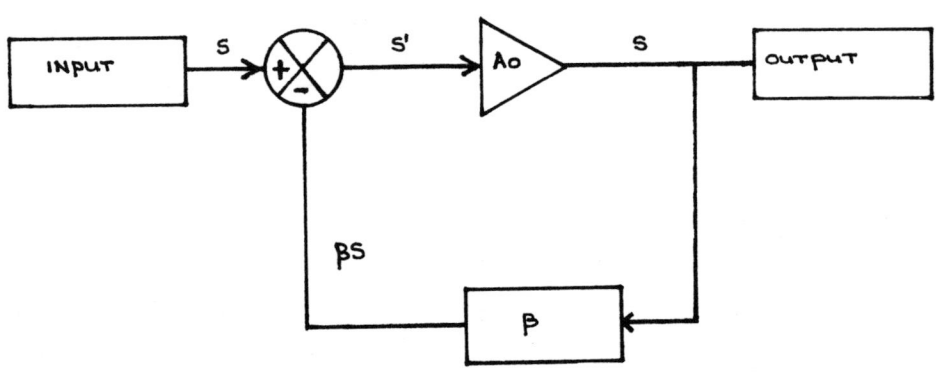

Figure 3.5 A feedback amplifier system

The circle with quadrants is a *mixer*. The signs in the quadrants indicate whether we add or take away the input to that quadrant when calculating the output of the mixer. The quantity, β, is the *feedback fraction*, which is the fraction of the output we take as a sample. Do not confuse the feedback fraction with current gain, which has the same symbol. The context is usually sufficient to indicate whether β is the feedback fraction or the current gain, but if there is cause for confusion you should use B instead of β.

The mixer output is: $\qquad\qquad s' = s - \beta S$ $\qquad\qquad$ (3.2)

The amplifier output is: $\qquad\qquad S = A_0 s'$ $\qquad\qquad$ (3.3)

Combining (3.2) & (3.3) gives:

$$S = A_0 s - A_0 \beta\, S$$

$\Rightarrow$

$$S = \frac{A_0 s}{1 + A_0 \beta} \qquad\qquad (3.4)$$

128

So the gain of the system, with feedback, is:

$$A' = A_0/(1 + A_0\beta)$$ (3.5)

A' is called the *closed loop gain* (this is why A_0 is the open loop gain). $A_0\beta$ is called the *loop gain*.

The Effect of Feedback on Input Impedance

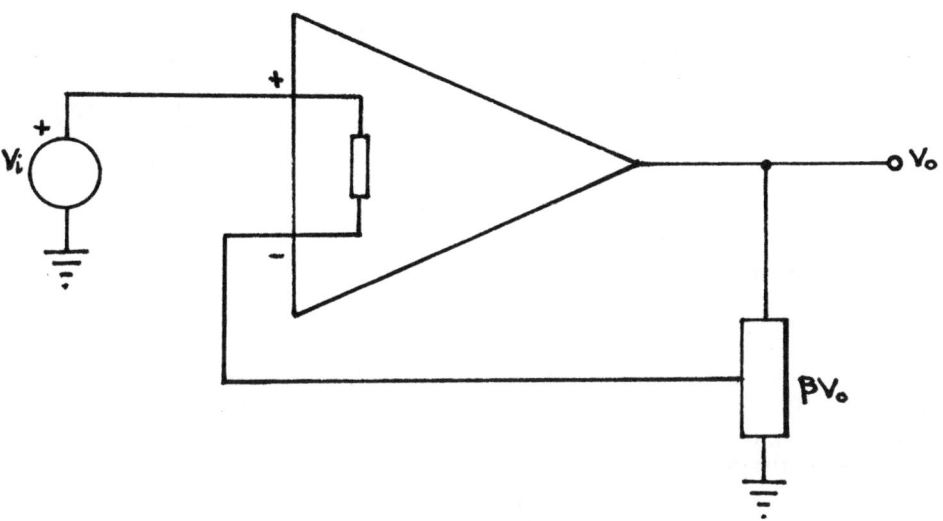

Figure 3.6 Finding the effect of feedback on input resistance

Suppose we apply feedback to a non-ideal amplifier which has a finite input impedance, R_i.

$$V_+ - V_- = V_i - \beta V_0$$

$$\Rightarrow \qquad I_i R_i = V_i \{ 1 - A_0\beta / (1 + A_0\beta) \}$$

$$= V_i / (1 + A_0\beta)$$

The input impedance, $R_i' = V_i/I_i$, is quite considerably improved by feedback.

$$\therefore \qquad R_i' = (1 + A_0\beta) R_i \gg R_i$$ (3.6)

The Effect of Feedback on Output Impedance

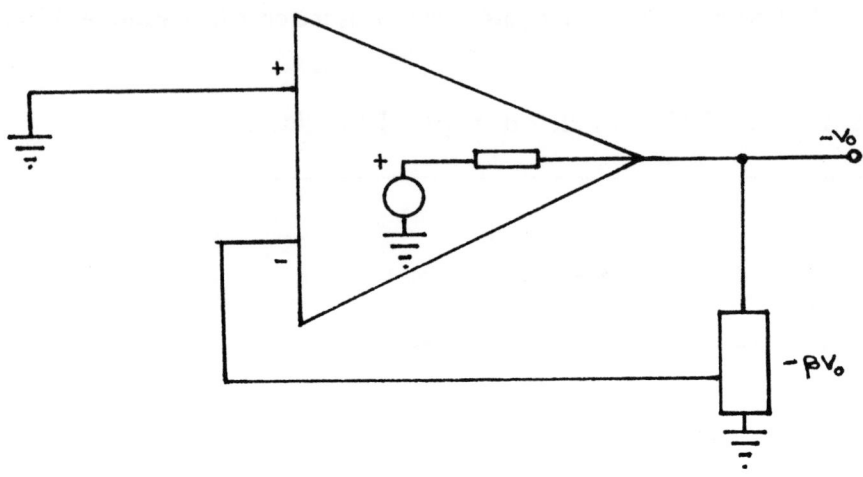

Figure 3.7 Finding the effect of feedback on output resistance

Here feedback is applied to a non-ideal amplifier which has a non-zero output impedance, R_o, and whose output e.m.f. is $E_o = A_0(V_+ - V_-)$. The output impedance can be simply calculated using a trick, which is to short circuit the non-inverting input and apply a voltage $-V_o$ to the output.

$$V_+ - V_- = \beta V_o$$

$$\therefore \qquad E_o = A_o \beta V_o$$

$$\Rightarrow \qquad I_o R_o = E_o + V_o$$

$$= V_o(1 + A_o \beta)$$

The output impedance is also improved by feedback.

$$\therefore \qquad R_o' = R_o/(1 + A_o\beta) \gg R_o \qquad (3.7)$$

3.1.3 Real Amplifiers

In a real amplifier A_0 will vary from device to device, and also with temperature, output, input, humidity, time for which it has been switched on, number of people looking at it, etc. This is regrettable but unavoidable, unless A_0 is made as big as possible.

Look at (3.5) in the condition that $A_0 \rightarrow \infty$

If $A_0\beta \gg 1$ $\qquad\qquad\qquad A' \approx A_0/A_0\beta$

$\Rightarrow$ $\qquad\qquad\qquad\qquad \lim_{A_0 \rightarrow \infty} A' = 1/\beta$ $\qquad\qquad\qquad$ (3.8)

The gain is now predictable, because β is likely to vary much less than A_0 will, and so the amplifier is very stable. The stability occurs because, although external factors may cause the open loop gain to vary by a hundred or so, the open loop gain is designed to be very high (~ 100 000) and these changes are insignificant.

Is it possible to make $A_0 \rightarrow \infty$? Yes, it is. Current source active loads, which have a very high impedance due to their current source behaviour, give very high gains (see long-tailed pair).

3.1.4 Operational Amplifiers

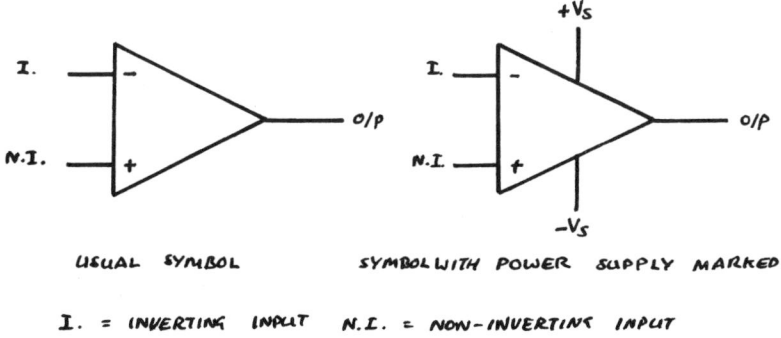

USUAL SYMBOL   SYMBOL WITH POWER SUPPLY MARKED

I. = INVERTING INPUT   N.I. = NON-INVERTING INPUT

Figure 3.8 Operational amplifier symbols

Operational amplifiers are integrated circuit amplifiers specifically designed to fill this role of high gain differential amplifiers precision systems. They are designed to match, as closely as possible, the properties of an ideal amplifier. Those properties are:

infinite gain;
infinite bandwidth (i.e. same gain at all frequencies);
infinite input impedances;

zero output impedance;

zero input-voltage offset (input offset is the p.d. required between the inputs to make the output zero); *and*

true differential amplification with infinite common mode rejection.

The universal symbol for an operational amplifier is that given in the previous examples.

Usually the power supply connections are omitted from the circuit symbol. In most applications the power supply is between $+V_{CC}$ and $-V_{CC}$, which means that the quiescent output voltage is 0V. In other cases the quiescent output voltage is half way between the voltage at the positive power supply terminal and that at the negative power supply terminal.

An operational amplifier obeys the ideal differential amplifier equation.

$$E_o = A_0(V_+ - V_-) \tag{3.9}$$

There are two assumptions that you can use which make analysis of amplifier circuits easier, and which remove the need to use (3.9).

They are:

The input terminals are always at the same potential; *and*

There is no current drawn at either input terminal.

The second statement is justified by the infinite input impedance of the circuit. The first is justified as follows:

The differential input voltage is given by a rearrangement of (3.9).

i.e. $$V_+ - V_- = V_o/A_0$$

V_{out} is finite and A_0 is infinite $\therefore$ $V_o/A_0 = 0$.

$\therefore$ $$V_+ = V_- \tag{3.10}$$

3.1.5 Real Operational Amplifiers

Amplifiers cannot be manufactured to match the specifications of the ideal amplifiers, but they can be manufactured so that the deviations are not significant in normal operation.

Finite Gain

Real operational amplifiers have a finite gain. It is, however, very high. Typical values are 100 dB (or 100 000) or higher.

Finite Bandwidth

Op-amps divide into two classes: those with internal frequency compensation and those without. By far the most popular op-amp, the 741, does, so I shall only discuss compensated devices here. The effect of frequency compensation is to make the open loop gain roll off at -20 dB/decade, starting at somewhere between 1 kHz and 10 kHz. This is done so that the amplifier has a low gain at any frequencies at which it might oscillate (see chapter 5).

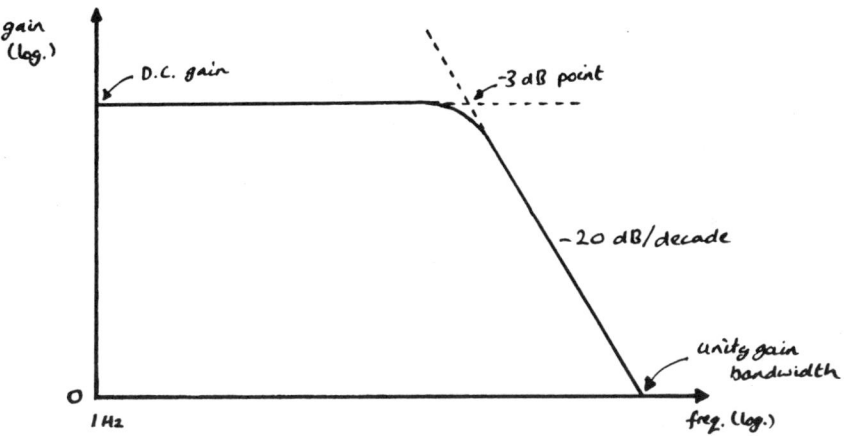

Figure 3.9 Operational amplifiers gain

In data sheets there are two items which are all that is required to find the open loop gain of most compensated op-amps at any frequency. They are the unity gain bandwidth and large signal voltage gain. Using these data you can plot a curve which gives the gain at any frequency.

Plot the frequency on the x-axis and the gain on the y-axis. You should use logarithmic scales or log-log graph paper. At the lowest frequency plotted (make this less than 1 Hz) mark the gain to be the large signal gain (d.c. gain). Then take a straight line across parallel to the x-axis and passing through this point. At the frequency equal to the unity gain bandwidth mark the gain to be 1 (0 dB). Many op-amps are designed so that the gain *rolls-off* nice and predictably at -20 dB/decade, and that corresponds to a gradient of -1 on a log-log graph. Thus you can draw the -20 dB/decade line back from the unity gain point until it meets the d.c. gain line. Mark the two lines in bold and you have the gain at all operating frequencies.

Using this graph, the approximate bandwidth at any frequency can be found. Since A' ≤ A$_o$ it is immediately obvious, by plotting the line "gain = A'", that the bandwidth of the amplifier is also increased by feedback.

133

Finite Input Impedance

The input impedance (between the inputs) is typically some MΩs for a bipolar operational amplifier. Unless the feedback loop is formed from components with similar impedances the current flowing into the inputs can be neglected, and so the infinite input impedance assumption is good.

Non-Zero Output Impedance

The output impedance of an operational amplifier is not zero. It is usually much lower than the components forming the feedback loop and so can be ignored. Errors will occur if you demand a high current source/sink capability from a standard operational amplifier, since it will not be capable of it.

There are two choices if you do wish to control high currents:

Use a *power operational amplifier*, specially designed to cope with high currents;

or   Attach a follower which has a high current capacity, and connect the feedback network to the output of the follower.

Finite CMRR

Commercially available op-amps have very high CMRRs (find this in a manufacturer's data sheet) and so the gain to common mode signals will usually be insignificant.

Non-Zero Input Offset

The input offsets are typically a few mV, or less, depending on the IC used. However, many ICs are provided with *offset null* terminals. In critical applications these are wired up, as recommended by the manufacturer, and adjustments made to force the input offset to zero.

Temperature Dependence

All the above properties do, in fact, have a temperature dependence. However, since the devices are manufactured to conform to an ideal in which all properties are either infinite or zero the temperature dependences are generally slight. It does not matter if the gain varies by 100 over a $50°$ operating range if the gain is 100 000 or more.

Slew Rate Limitations

The slew rate is the maximum rate at which the output of an operational amplifier can change. It is a property of all real amps and causes a deterioration of *fast edges* in pulse.

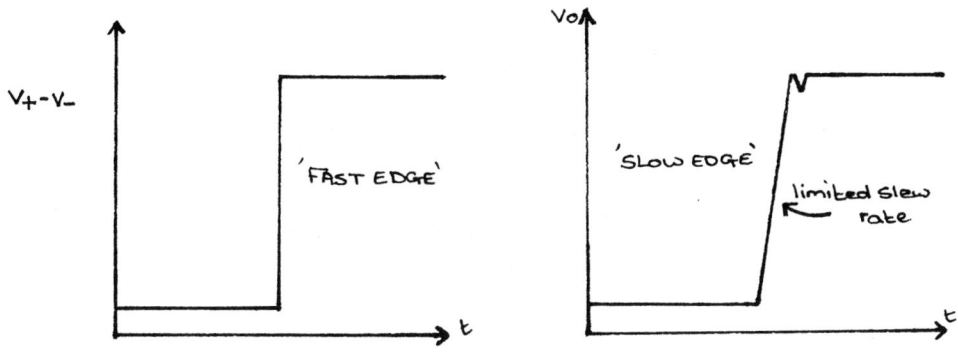

Figure 3.10 Effect of finite slew rate

Manufacturers list many other properties.

3.2 Operational Amplifiers As Precision Controllers

This section shows how operational amplifiers can be used in *control* applications. The applications discussed here are precision voltage generators and current sources.

The voltage generators are very important circuits. They are used for *regulated power supplies*, which are power supplies which appear to have very low internal resistance. In reality, this section ought to be called 'voltage regulators, current sources, and power supplies' and included in a separate chapter. I am including it here because it is a major application of system stabilisation via feedback and the use of operational amplifiers.

3.2.1 Precision Voltage Generators

To build a voltage generator which appears to have zero internal resistance we require a circuit similar to the platform controller discussed in §3.1. The output voltage is compared with a stable reference voltage, and modified when there is error. The following, simple feedback amplifier, circuit works.

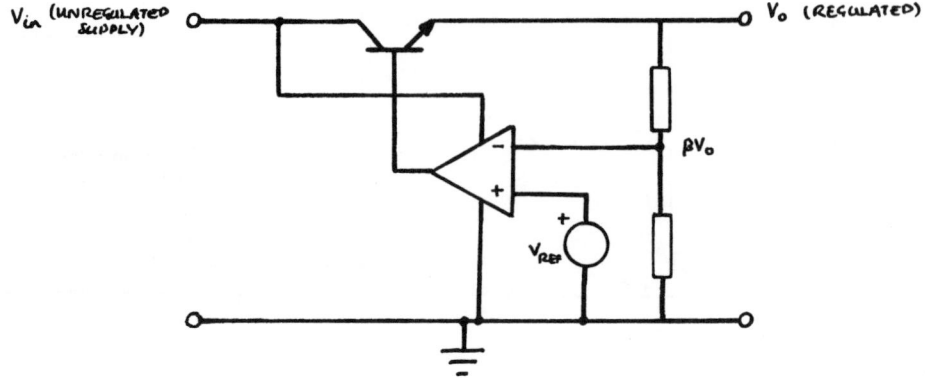

Figure 3.11 Operational amplifier voltage regulator

The analysis is simple. The operational amplifier has the positive power supply taken from the input positive line and the negative supply is tied to earth (0 V). The quiescent output is $\frac{1}{2}V_{in}$ and the amplifier output is at all times:

$$V = \tfrac{1}{2}V_{in} + A_0(V_+ - V_-) \qquad (3.11)$$

V_+ is determined by the reference voltage, which is precise and stable, and V_- is a fraction of the output of the circuit, βV_0. Finally, there is a diode drop between the amplifier output and the circuit output. i.e. $V_0 = V - V_{BE}$. Substituting all this into (3.11):

$\Rightarrow$ $\qquad\qquad\qquad V_0 = \tfrac{1}{2}V_{in} - A_0(V_{ref} - \beta V_0) - V_{BE}$

$\Rightarrow$ $\qquad\qquad\qquad V_0(1 - A_0\beta) = \tfrac{1}{2}V_{in} - A_0 V_{ref} - V_{BE}$

$$V_0 = \frac{\tfrac{1}{2}V_{in} - A_0 V_{ref} - V_{BE}}{(1 - A_0\beta)} \qquad (3.12)$$

The term $A_0\beta$ occurs in many operational amplifier formulae. It is refered to as the *loop gain*. Throughout the preceding analysis I have used the assumption that the input impedance of the amplifier appears infinite, and the output impedance appears zero, in comparison to other circuit elements.

In the limit $A_0 \rightarrow \infty$ $\qquad\qquad V_{out} = V_{ref}/\beta \qquad (3.13)$

To demonstrate the use of the simplifying assumptions discussed in §3.1 I shall arrive at (3.13) in two lines:

$$V_+ = V_- \ \Rightarrow \ V_{ref} = \beta V_{out}$$

$$\Rightarrow \qquad\qquad V_{out} = V_{ref}/\beta$$

That was easy wasn't it! High currents can be passed by the transistor, but the output voltage will not be significantly affected by the load current drawn. The transistor to use is a 2N3055, which is a favourite for power supply applications. It can conduct a maximum collector current of 15 A at a maximum power dissipation of 115 W!

In a real circuit the output is a function of the input, and if the power supply to the amplifier has a 100 Hz ripple the output of the amplifier will have a 100 Hz ripple. The magnitude of this effect depends on a factor called the *power supply rejection ratio*. The power supply rejection ratio, or PSRR, is the ratio differential-gain:power-supply-gain, and the bigger it is the less ripple will emerge at the output. For a *sensitive* application choose an operational amplifier with large PSRR.

Operational amplifier non-ideality affects the performance of such a *voltage regulator* circuit, but the largest source of error could be due to the reference voltage. However, you should not get the impression that op-amp voltage regulators are unsatisfactory - negative feedback is so powerful that a circuit built with standard, *off the shelf*, components is usually a very good performer.

3.2.2 Op-amp Stabilised Current Sources

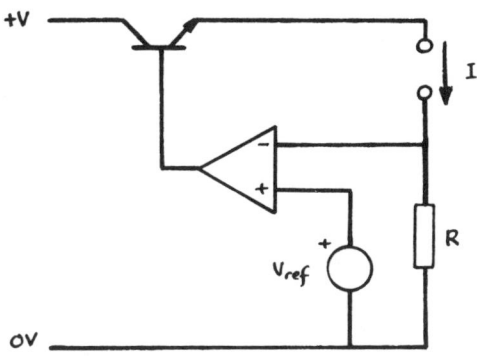

Figure 3.12 Operational amplifier current source

In the same way as the feedback stabilises a voltage generator, it can also be used to stabilise a current source.

From the ideal amplifier assumptions:

$$V_+ = V_- \implies V_{ref} = IR \implies I = V_{ref}/R \qquad (3.14)$$

3.2.3 Stable Voltage References

Both the above circuits are good circuits in that the feedback provides stability. However, Zener diode references are not the most accurate, or stable, available. This section discusses making accurate and stable voltage references, whilst the next section shows how the principles discussed are applied in integrated circuit regulators, and how to use those devices.

Zener Diode Regulators

The simplest reference is the Zener diode, which is cheap and usually quite reliable. However, breakdown voltages are usually not guaranteed, by the manufacturer, to within less than 5%.

IC References

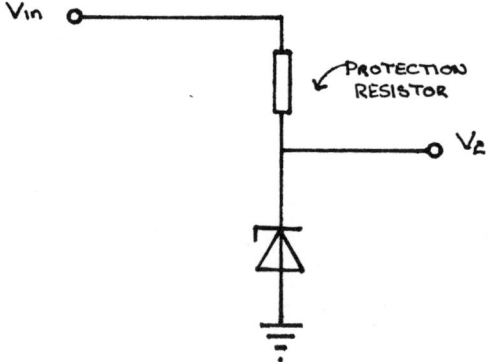

Figure 3.13 Zener diode voltage reference

Many integrated circuit voltage references are based on the *bandgap* reference. This is a clever little circuit in which the negative temperature coefficient of the base-emitter junction is eliminated by generating a voltage with an identical positive temperature coefficient. Prices go up to about £5 or more. An example is the REF-02, which comes in a 8-pin round can, and outputs 5V.

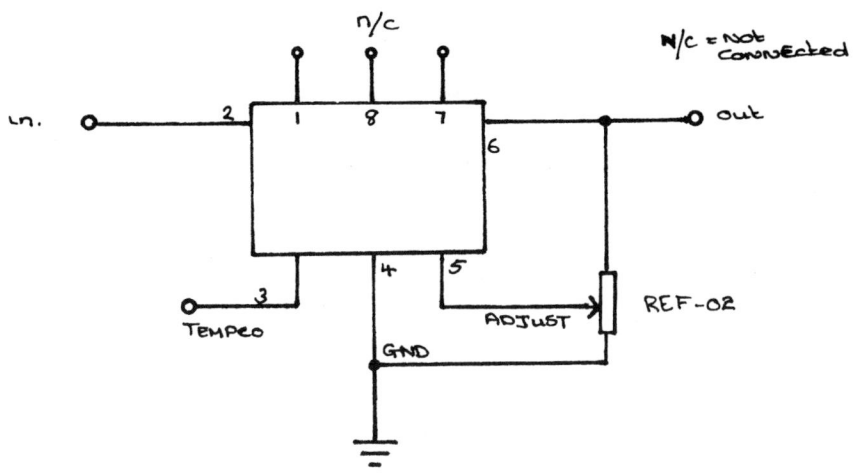

Figure 3.14 REF-02 voltage reference

It has an excellent temperature coefficient of 20 ppm/°C (ppm = parts per million = 0.0001%), and an adjust pin for trimming to a precise value, line regulation of 0.009%/V, load regulation of 0.006%/mA, etc. It costs (1988) £4.20.

A cheaper alternative is the TL430C/TL431C *programmable zener diode*. This is a circuit which uses the transistor V_{BE} to generate a reference. It has a temperature coefficient (tempco) of 10 ppm/°C and is fully programmable, by an external pot., to voltages ranging from about 3 - 30V.

3.2.4 IC Regulators

Three Terminal Regulators I

For non-critical applications you can buy the whole regulator circuit, complete with reference voltage, differential amplifier, series pass transistor and feedback loop, in a three pin IC package.

The simplest of these are probably the 7800 series of regulators. The voltage is set by the manufacturer (and represented as the last two digits of the device's number, e.g. 7806 gives a 6V output).

This provides good regulation (consult a data sheet for details) at low currents, and at a low price. To boost the current you can buy higher power versions, or you can add a power transistor.

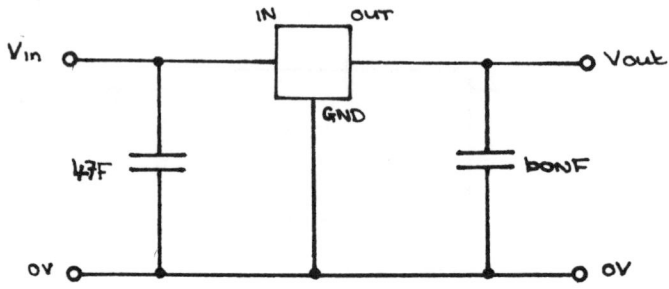

Figure 3.15 Simple 3-terminal regulator

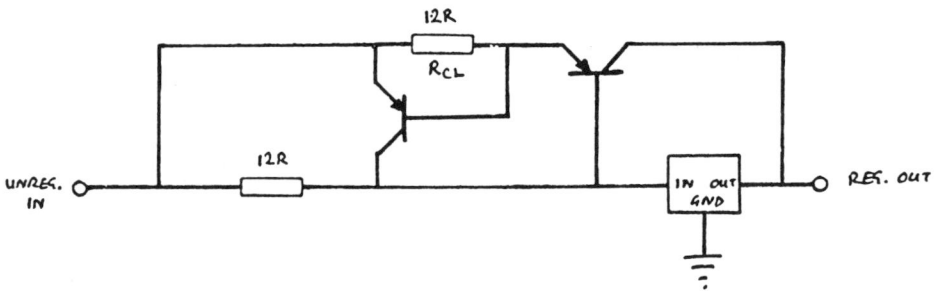

Figure 3.16 High current regulator

When the current is below 50 mA the transistor is off and the circuit acts as before. When the current exceeds 50 mA the transistor is turned on and conducts a high current. The IC, however, still keeps the output at the regulated voltage and you have a high current regulator. The input must be kept above the output by the *dropout voltage*, which is required so that the circuit actually works, plus a diode drop for the transistor.

One of the things you learn from experience is that your expensive power transistor sometimes blows quickly, and protects your cheap fuse from damage. It is better to protect the circuit with electronic current limiting.

When about 0.6 V is dropped across R_{CL} the extra transistor turns on and robs current from the 12 Ω resistor. This reduces the current passed by the original transistor and prevents it being destroyed in a short circuit condition.

140

Three Terminal Regulators II

For a more versatile regulator you can use a device such as the LM317. This has an adjust terminal in place of the ground terminal, and the feedback network is formed externally to the device.

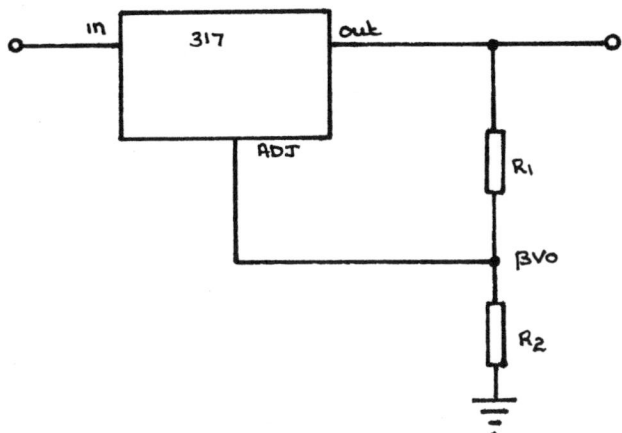

Figure 3.17 317 Regulator circuit

The device acts to keep the potential between the output and adjust pins equal to 1.25 V.

i.e. $$V_{oa} = V_o - V_a = 1.25 \text{ V}$$

if $V_a = \beta V_o \Rightarrow$ $$V_o = V_{oa}/(1 - \beta)$$

In the circuit $\beta = R_2/(R_1 + R_2)$

$\therefore$ $$1 - \beta = R_1/(R_1 + R_2)$$

$\Rightarrow$ $$V_o = 1.25(1 + R_2/R_1) \tag{3.15}$$

Complete Power Supply Regulator Circuits (723 Regulators)

This subsection deals with the μA723 regulator circuit. Essentially, it is an integrated version of the circuit discussed in §3.2.1. However, unlike the three terminal regulators, you have to link up the operational amplifier, voltage reference, feedback loop, etc., externally and so the device is totally flexible. It comes in a round, ten pin, metal can or a 14-pin plastic package.

141

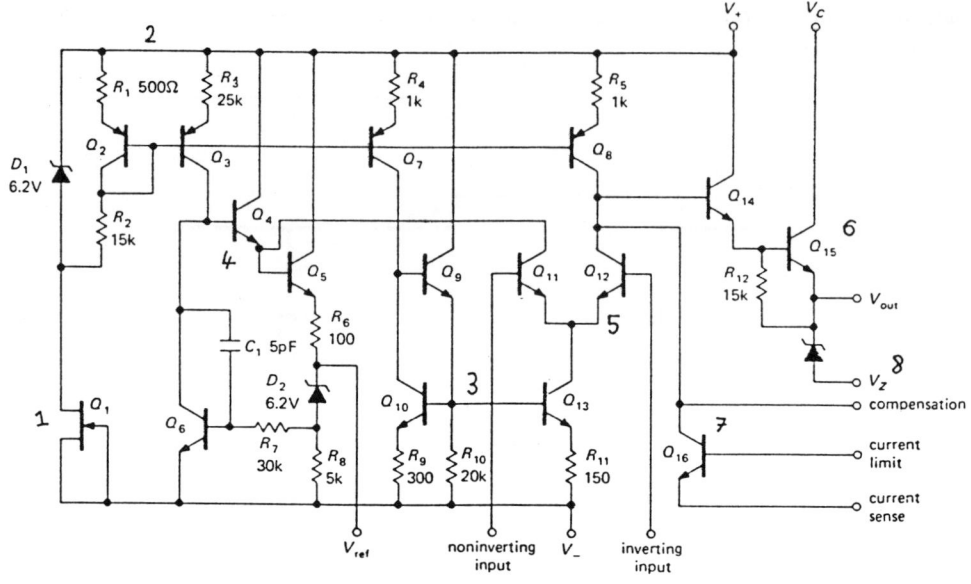

Figure 3.18 Schematic of 723C regulator (Courtesy of National Semiconductor)

The circuit is what is inside the package. Although it looks complicated, it is quite a simple circuit. The highlighted areas are:

1   FET current source;
2   current mirror - the resistors alter the current ratios;
3   Wilson current mirror;
4   high gain amplifier;
5   differential amplifier;
6   higher power output stage; and
7,8   protection circuits to allow current limiting and overvoltage shutdown.

The messy-looking circuit on the LHS is a temperature compensated, current source driven (1), zener diode voltage reference with an Darlington emitter follower (4) output stage to remove loading effects on the zeners. The reference (nominally 7.15V) is brought out to the pin V_{ref}.

The RHS is a simpler circuit. The Wilson mirror (3) provides the emitter current source for the differential amplifier (5). The other mirror (2) is used to produce current source active loads for the amplifier (5). This gives a very high gain differential amplifier with both inputs brought out to pins, and external frequency compensation for maximum flexibility.

The output stage (6) has its own power supply, so the regulated supply can be completely separate from the amplifier supply. In very high power applications (6) can be connected, as the low power end of a Darlington, to a transistor such as a 2N3055.

The current limit circuitry is simple. When ≈ 0.6 V is applied across the base of the transistor (7) it robs base current from the output stage (6), and thus limits the output current.

The overvoltage protection (8) is a feature only available on the 14-pin plastic package. If, for some reason, the output voltage is too high the zener diode starts to conduct. If the V_Z terminal is connected to an SCR crowbar circuit the regulator and load are immediately shorted out, blowing the fuse and protecting voltage sensitive circuits, such as microprocessors.

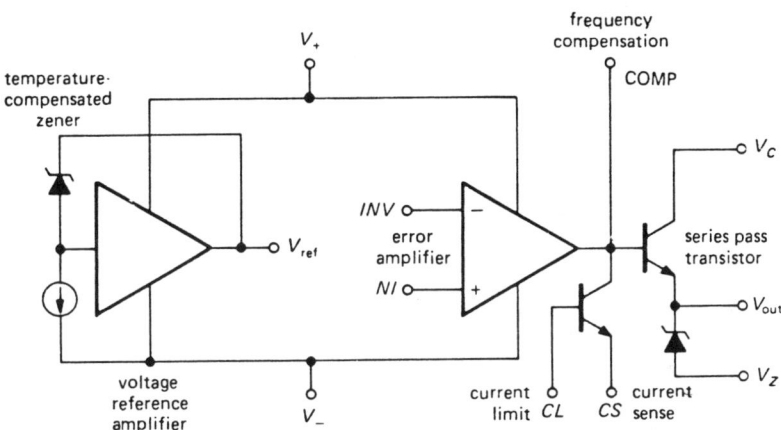

Figure 3.19 Simplified schematic of 723C (Courtesy of National Semiconductor)

Figure 3.19 is a symbolic representation of what the the circuit actually does. Using this, and the circuit of §3.2.1, you should find it very easy to design power supply regulator circuits around the device.

Low Power 723 Precision Regulator

A basic regulator is shown in Figure 3.20.

For best operation of the differential amplifier the circuit should see the same impedance connected between the inverting input and ground as is between the non-inverting input and ground. Hence the inclusion of R_{NI}, for which:

$$R_{NI} \approx R_{I1} || R_{I2}$$

143

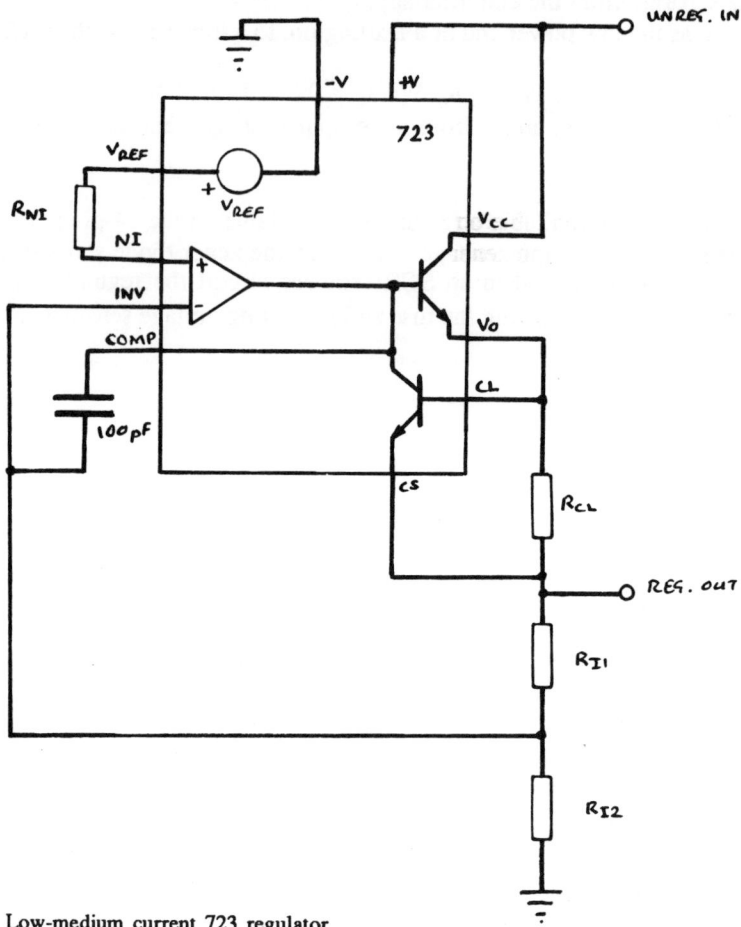

Figure 3.20 Low-medium current 723 regulator

The 100 pF capacitor, wired between the inverting input and COMP pin, acts to give the amplifier frequency compensation as discussed above. The COMP pin is wired to the amplifier output. You should ensure that the current limiting resistor, R_{CL}, is rated for the high powers it will have to dissipate. Also, the value of V_{REF} actually lies between 6.8 & 7.5 volts so, to ensure an accurate design, it might be a good idea to have an adjustable feedback network.

High Power 723 Regulator

This circuit is basically the same, although there are a few extra points to note. The current limiting transistor will turn on at about 0.5 V (not 0.6-0.7) if the regulator is run at a high power and is hot. For high powers you should get the metal can version of the 723, and mount it on a heatsink. The maximum current that the 723's output stage may conduct is 150 mA, and the maximum power dissipation of the package is 1 watt. Ensure that your parallel pass transistor can conduct short circuit currents and can handle the power dissipated when not only does it conduct short circuit currents, but it has the entire unregulated supply p.d. dropped across it.

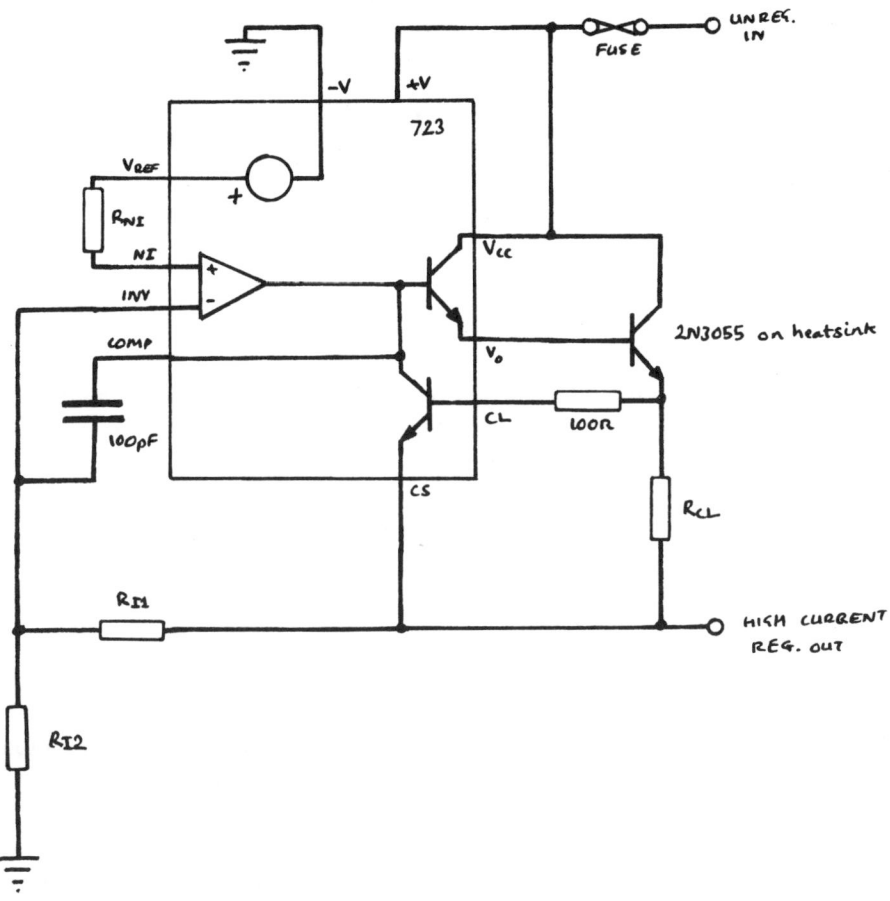

Figure 3.21 High current 723 regulator

Parallel Parallel Pass Transistors

If you find that the power dissipated by your parallel pass transistor will result in junction damage in short circuit conditions (junction temperature cannot exceed 200°C for metal can devices) you always have the option of splitting the current by connecting identical transistors in parallel.

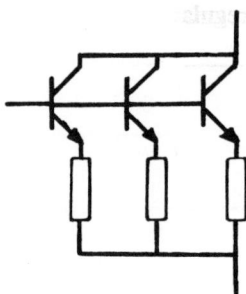

Figure 3.22 Parallel pass transistors

The resistors are added because the devices will not be the same, and will have slightly different V_{BE}s at any given current, and so one device might end up conducting more current than it is supposed to. The value of the resistors is chosen to give approximately 0.2 V drop at maximum output current.

This method can also be used to reduce cost. It might be cheaper to buy five 2 A transistors rather than one 10 A transistor.

Foldback Current Limits

This special type of circuit reduces the need for components to withstand continuous short circuit conditions. The circuit detects if there is a short across the output and limits the short circuit current to *less than the maximum output current*.

Under maximum output conditions, 2V is dropped by the 1 Ω current limit resistor. Because of the pot. the current limit transistor's base is held at 15.5 V (0.9 ×17 V) and so there is 0.5 V across the base-emitter junction of the limit transistor. If the output current increases slightly the limit transistor will start to rob current, limiting the output source capability. In a short circuit condition the limit transistor's emitter is tied to 0 V. The current is still limited to that which causes a 0.5 V drop across the base-emitter junction, and so the pd. across the current limit resistor is 0.5/0.9 or 0.6 V. The short circuit output current is 600 mA, not 2 A.

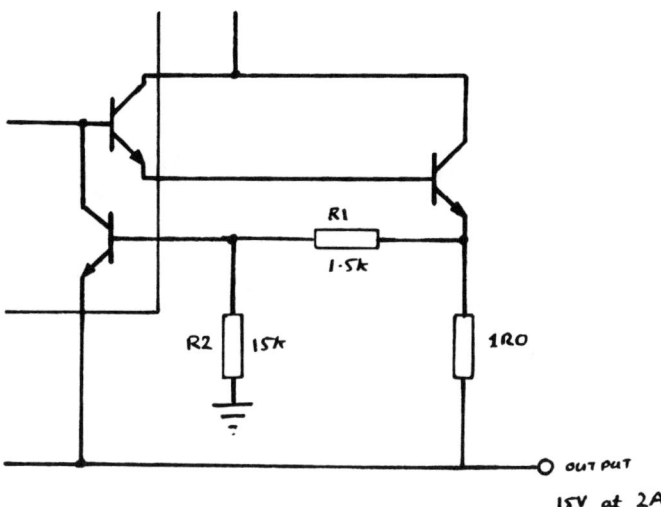

Figure 3.23 Foldback current limit circuit

3.3 Feedback Amplifiers and Other Operational Amplifier Circuits

If you have wired up at least one of the circuits in §3.2 you should have been impressed by the ability of the feedback circuit to stabilise the output of the voltage regulator circuit. If, for any possible reason (and this includes noise, temperature effects, loading effects, etc.), the output of the amplifier deviates from the desired output of V_{REF}/β the circuit takes action to bring the output back to what it should be.

Suppose that V_{REF} were replaced with a variable voltage generator, such as a stylus, a microphone, a tape head, a light meter, etc. The result would be a stable, low noise, amplifier. Such a *feedback amplifier* is the first subject of this section, but the near-ideal properties of the operational amplifier make it a very versatile device and I shall go on to consider some of the many applications of the operational amplifier.

3.3.1 An Operational Amplifier as a Non-Inverting Feedback Amplifier

The feedback amplifier can be easily wired up using an operational amplifier. For an ideal amplifier we would expect its gain to be given by $1/\beta$.

i.e. $$A' = 1/\{R_{F2}/(R_{F1} + R_{F2})\}$$

or $$A' = 1 + R_{F1}/R_{F2} \qquad (3.16)$$

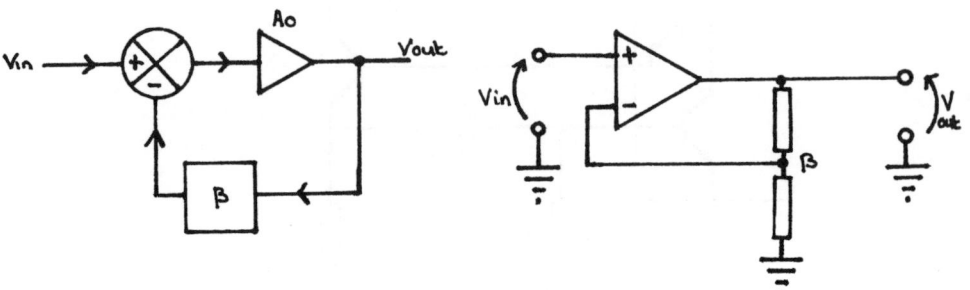

Figure. 3.24 Connecting an op-amp as a 'classic' voltage feedback amplifier

---

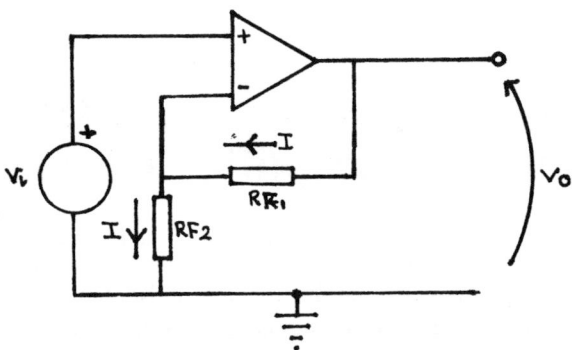

Figure. 3.25 Circuit to calculate gain of non-inverting feedback amplifier

---

Since we are dealing with an ideal amplifier, I shall analyse it with the two simplifying assumptions.

The input to the circuit is V_i, and because of the infinite gain $V_+ = V_- = V_i$. The current flowing through R_{F2} is I.

$$I = V_i/R_{F2}$$

Because of the infinite input impedance this current cannot be sourced by the inverting input, and so must be sourced by the output. This means that the same current must be

148

flowing through R_{F1} as through R_{F2}. If the current I flows from the output to ground through the feedback loop then the output will rise to a potential, V_o, above ground.

$$V_o = I(R_{F1} + R_{F2})$$

therefore:
$$V_o = V_i \times (1 + R_{F1}/R_{F2}) \qquad (3.17)$$

Comparing (3.17) with (3.16) it can be seen that the gain is indeed $1/\beta$, which is reassuring.

Also the circuit has impedances:

$$r_{in} = \infty \quad \& \quad r_{out} = 0$$

This will not quite be true for a real circuit, but it will be sufficiently accurate for most cases.

If we let $R_{F2} \to \infty$ and $R_{F1} \to 0$ we get:

$$V_o = V_i$$

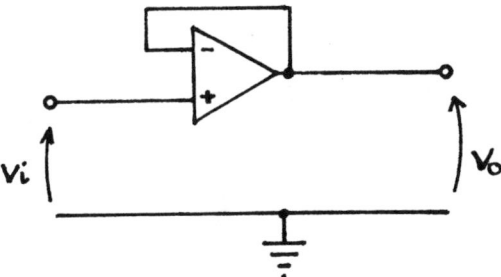

Figure 3.26 Operational amplifier voltage follower

The ideal follower!

3.3.2 Other Operational Amplifier Circuits

Inverting Feedback Amplifier

The non-inverting amplifier is almost the ideal amplifier. It has well defined gain, infinite input impedance and zero output impedance. But no matter how hard you try, you can never make the gain negative! With the following circuit you do.

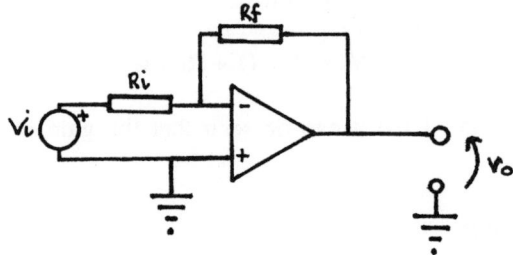

Figure 3.27 Inverting amplifier

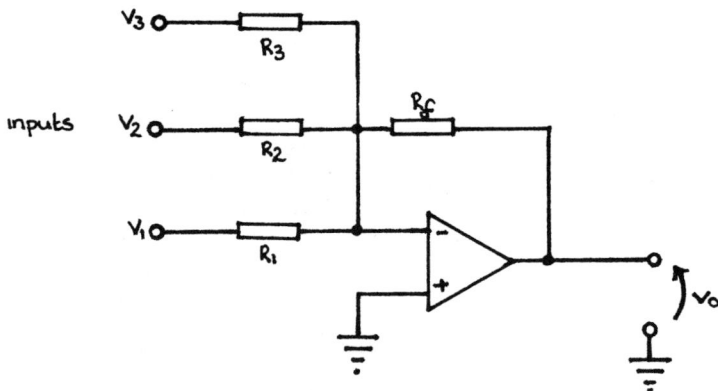

Figure 3.28 Summing amplifier

The analysis is again done using the two simplifying assumptions. It is from this circuit that the assumptions get their name - the virtual earth principle. The inverting input must be at zero volts, so the current flowing through the feedback loop is:

$$I = V_i/R_i$$

Consequently, the p.d. dropped by R_f is:

$$V_{Rf} = IR_f = V_iR_f/R_i$$

But, since both inputs are at zero volts:

$$V_o = - V_i \times R_f/R_i \qquad (3.18)$$

The gain has been made negative but we do pay a price for this. The input impedance, with an ideal amplifier, is no longer infinite!

$$R_{in} = R_i \quad \& \quad R_{out} = 0$$

Summing Amplifier

This is a simple variation on the previous circuit. From the virtual earth principle it is easy to show that:

$$V_{out} = - R_f \sum V_i/R_i \qquad (3.19)$$

Differential Amplifier

A differential amplifier with a well defined finite gain is often useful.

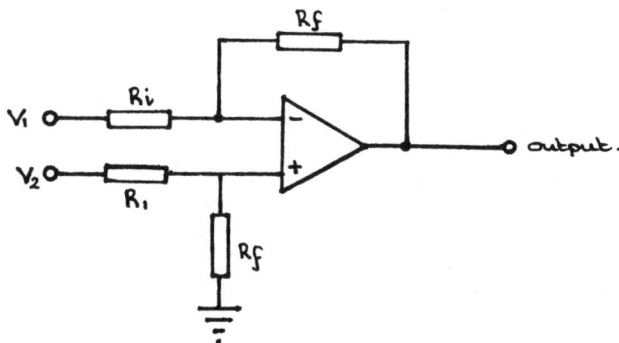

Figure 3.29 Differential amplifier

From the virtual earth principle:

$$V_+ = V_2 \times R_f/(R_f + R_i) \qquad \text{(a)}$$

&
$$(V_1 - V_-)/R_i = (V_- - V_o)/R_f \qquad \text{(b)}$$

(b) rearranges (eventually) to give (c):

$$V_o = -\frac{R_f}{R_i}\left(V_1 - \frac{R_f + R_i}{R_f} V_-\right) \hspace{2cm} \text{(c)}$$

But $V_- = V_+ \Rightarrow$ $$V_o = -\frac{R_f}{R_i}(V_1 - V_2) \hspace{2cm} (3.20)$$

3.3.3 Integrators, Differentiators & Analogue Computation

The circuits in this section act as the *differential operators* d/dt and $\int$ dt on their inputs. Unlike with the simple RC integrators discussed in chapter 1, there are no approximations made and the calculus is exact.

Integrator

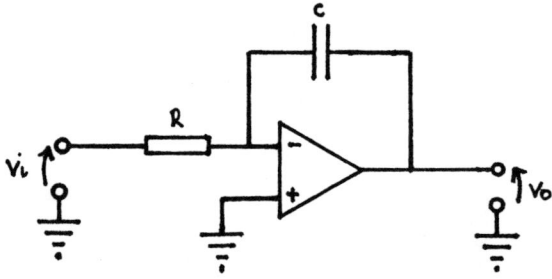

Figure. 3.30 Integrator

From the virtual earth principle, at any time:

$$dQ/dt = V_i/R$$

The capacitor is charged by a constant current, and we may write:

$$Q = CV_C \quad \text{or} \quad Q = -CV_o \Rightarrow dQ/dt = -C\, dV_o/dt$$

$$- V_i/RC = dV_o/dt$$

$$\Rightarrow \qquad V_0 = -\frac{1}{RC} \int V_i \, dt$$

Differentiator

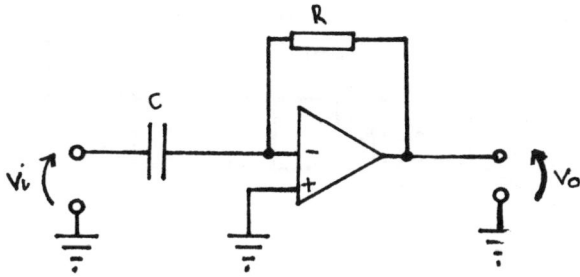

Figure. 3.31 Differentiator

$$dQ/dt = - V_o/R \quad \& \quad dQ/dt = C \, dV_i/dt$$

$$\Rightarrow \qquad V_o = - RC \, dV_i/dt$$

Analogue Computation

The above two circuits allow us to electronically solve differential equations. This is called analogue computation.

Example

Solve, subject to the given initial conditions, the equation:

$$d^2V/dt^2 = - \omega^2 V, \quad V(0) = 2 \text{ V}, \quad V'(0) = 0 \text{ V}$$

Solution

integration gives $$V = - \omega^2 \iint V \, dt^2$$

The initial condition is made by having all the capacitors initially discharged and then momentarily connecting the output to the 2 V pot. The solution to such an equation is an harmonic oscillation with an angular frequency of ω - i.e. a sinusoid.

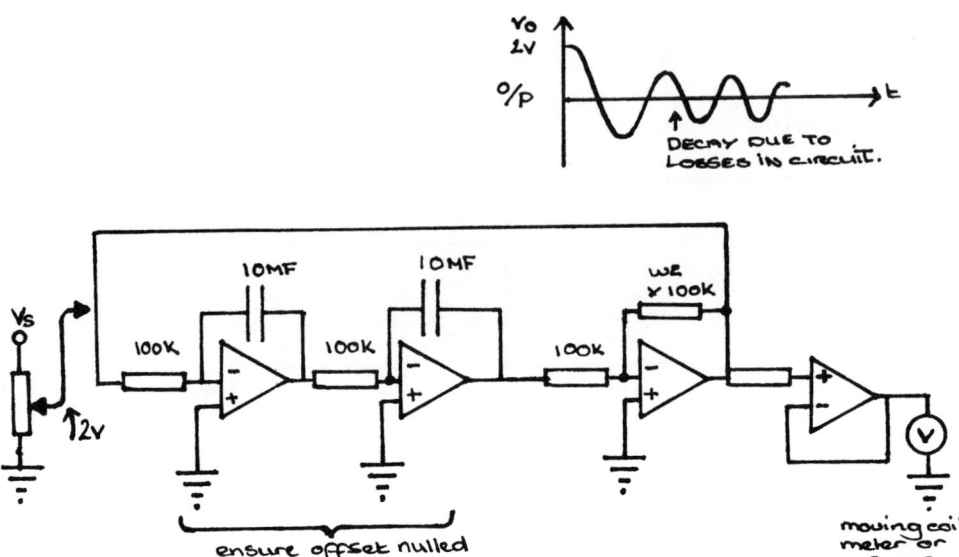

Figure. 3.32 Analogue computer to solve SHM equation

3.3.4 Some Other Circuits

Active Rectifier

This circuit performs a linear rectification with no diode drop.

Comparator

This circuit is a decision maker. If $V_+ > V_-$ the output goes to the positive supply rail (positive saturation), and if $V_+ < V_-$ the output goes to the negative supply rail (negative saturation).

The output can be either 0 V or V_S, because such circuits are usually powered from single supplies. The interpretation of the output is that if the output is V_S it is true that $V_+ > V_-$, and if the output is 0 V it is false that $V_+ > V_-$. An output which can take one of a set of discrete states is said to be *digital*. Digital circuits are discussed in chapter 4.

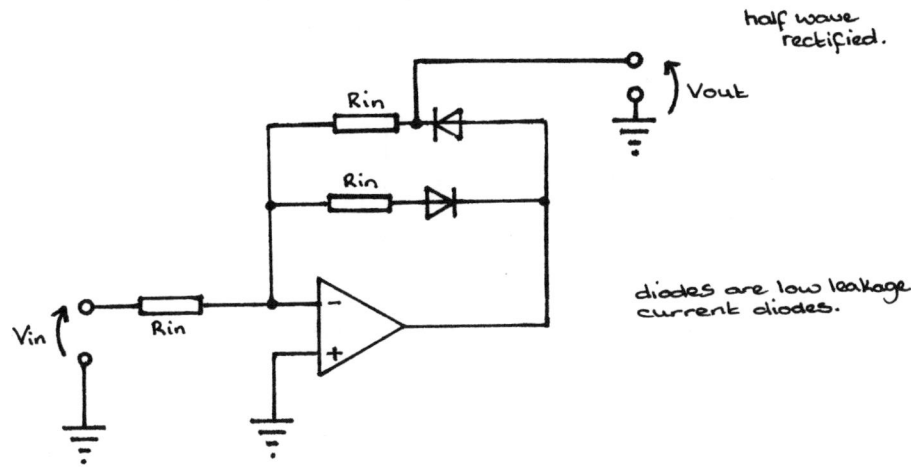

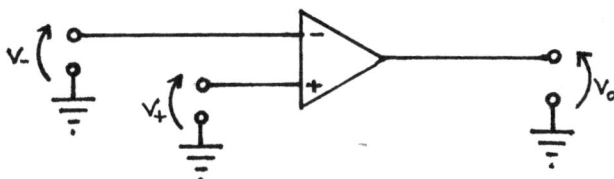

Schmidt Triggers

The Schmidt Trigger is a circuit designed to clean up digital data, which has gathered noise by transmission over a long distance. In such a situation we may wish to 'clean' the input to the receiver, and if we use a simple comparator we remove much noise, but generate errors.

The Schmidt Trigger removes the errors, caused by the noisy input, by altering the threshold level according to the state at the output. If the output is high, then the input must be greater than a certain value (the *high threshold*) to cause the input to go low, but if the output is low, the input must be lower than the high threshold by a preset amount before the output goes high. The preset amount is referred to as the *hysteresis* of the amplifier.

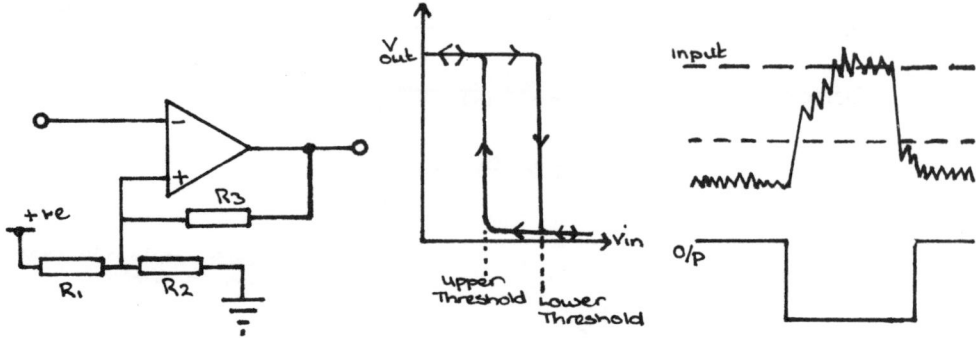

Figure 3.35 Op-amp Schmidt trigger, characteristic, and application

The operation is quite simple. With the output low, the upper threshold voltage is determined by the pot formed by R_1 & $R_2 || R_3$. With the output high, the lower threshold voltage is determined by the pot. formed by R_1 & $R_2 || R_3$. Because of their frequent use Schmidt Triggers are available in integrated circuit form and have a special symbol.

Questions for Chapter 3

3.1 An amplifier with gain 4 is set up to amplify voltages. If a 1 volt input is applied, calculate the percentage change in the output if the gain drifts by 5%.

3.2 An amplifier with open loop gain 60 dB is set up to perform the same task. Calculate the value of the feedback fraction to reduce the gain to 4, and the percentage change in the output if the open loop gain drifts by 5%.

3.3 A long tailed pair with insignificant common mode gain, 60 dB differential mode gain, 100 kΩ input resistance, and 2.2 kΩ output resistance is wired to act as a feedback amplifier of closed loop gain 100. Find the feedback fraction, the input resistance, and the output resistance, of the amplifier.

3.4 A differential voltage amplifier is constructed from a long tailed pair with significant common mode gain. Does the application of negative voltage feedback reduce the 'error' caused by the common mode gain? If so, by what factor?

3.5 A similar amplifier is constructed from a long tailed pair with negligible common mode gain but with significant second order gain*. Does the application of negative feedback reduce the error in this case? If so, by what factor?

*second order gain means $V_{out} = A_1(V_+ - V_-) + A_2(V_+ - V_-)^2$

3.6   Using op-amps and commonly available components design a voltage regulator to give a maximum output of 2 A at 15 V.

3.7   Design a ±15 V *dual tracking regulator*. This is a circuit which provides a +15 V regulated output, a 0 V output, and -15 V regulated output.

3.8   It is possible to control very high currents by adding output power transistors to the basic voltage regulator circuit. Suggest a method which could be used to produce a high voltage regulated supply - such as 500 V d.c. for example.

3.9   Figure 3.i shows a simple logarithmic amplifier. Show that its output can be written:

$$V_o = V_i + a \log bV_i$$

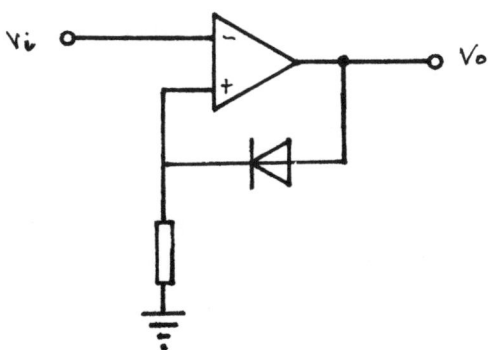

Figure 3.i A Logarithmic amplifier

Design an amplifier which gives a 'true' logarithmic output, unlike that of fig. 3.i.

Answers to Numerical Problems

3.1   5%
3.2   0.249; 0.02%
3.3 9×10^{-3}; 1 MΩ; 220 Ω

4

DIGITAL ELECTRONICS

4.1 Logic Circuits and Boolean Algebra

4.1.1 Electronic Decision Making

In the last chapter I introduced the comparator, a circuit which makes a true/false decision about a certain statement.

e.g. Is it hot?

When dealing with analogue systems this question can only be answered in terms of a position in a continuum.

e.g. It is $35.243534°C \pm 2 \times 10^{-6} °C$.

Such an answer is often not suitable. If a fan were to be turned on when it is too hot, and the designer decides that "too hot" means a temperature greater than 30°C, then the answer 35.24... is nonsense. The comparator is used to make the decision "it is too hot", and only has two possible outputs: i) it is TRUE that it is too hot; and ii) it is NOT TRUE (FALSE) that it is too hot.

However, most decisions which need to be made electronically are not as simple as the above example.

e.g. Is it true that the temperature is between 30°C & 35°C?

As it stands, we cannot answer this question electronically. However, if we rephrase it:

Is it TRUE that the temperature is GREATER THAN 30°C?

and Is it also TRUE that the temperature is LESS THAN 35°C?

we are in a position to answer the question. How is this possible? The state TRUE is defined to be represented by some output from a circuit, e.g. 5 V, and the state FALSE as some other output, e.g. 0 V. It is then very simple to design a transistor circuit which gives a 5 V (TRUE) output if, and only if, it has 5 V applied to both of its two inputs. This circuit is called a *logic gate* because it is making what is called a logical decision.

i.e.   Is it true that the input A is at 5 V AND the input B is at 5 V?

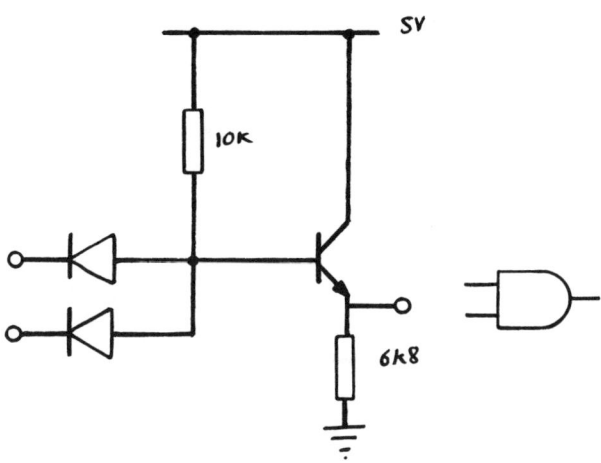

Figure 4.1 AND gate circuit and symbol

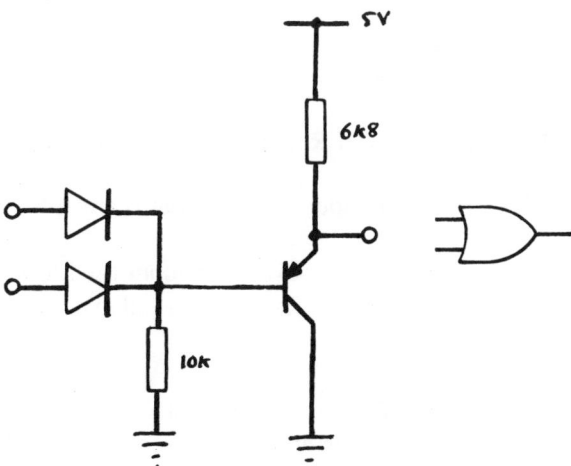

Figure 4.2 OR gate and symbol

Figure 4.1 shows a simple circuit which performs such a task. Its output is TRUE when the state at input A is TRUE AND the state at input B is TRUE.

Circuits can also be made to answer the following questions:

Is it TRUE that EITHER input A OR input B OR BOTH A AND B are at 5V?

This is called an INCLUSIVE OR gate (or just simply an OR gate) because it includes the statement OR BOTH A AND B.

Is it TRUE that the input is NOT at 5 V?

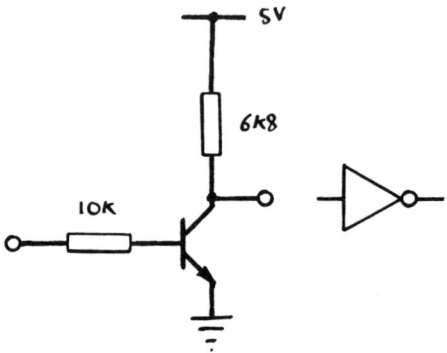

Figure 4.3 NOT gate and symbol

This is called a NOT gate or an INVERTER. The inverter is useful because it allows questions such as the examples below to be asked:

Is it TRUE that NEITHER input A NOR input B are at 5 V?

Is it NOT TRUE that input A AND input B are at 5 V?

Such circuits as the transistor circuits above are seldom used to make such decisions, since vastly superior integrated circuit versions are available. These ICs often contain four or more separate gates.

It is possible to express complicated decisions in terms of simple statements which can be processed by the electronic circuits above, but before you sit down and wire up a circuit with logic gates (which you have to buy) it would be nice to know if your design is, in fact, answering the correct question. What you need is a system with which you can analyse a logic circuit on paper to be sure that you have a correct answer (in the simplest possible way).

4.1.2 Boolean Algebra

Such analysis is possible using a type of mathematics originally developed to analyse logical problems. It is called *Boolean Algebra* and is very simple to use.

In algebra there are three basic ideas:

constants	which are numbers;
variables	symbols which can represent the value of any constant; *and*
operations	such as addition, multiplication, square root, etc.

In Boolean algebra there are only two constants, TRUE and FALSE. It is customary to represent these states with binary digits (*bits*), but some manufacturers prefer the less ambiguous symbols T and F.

Mostly we use: 1 for TRUE; *and*
                  0 for FALSE.

This is called *positive logic*. Negative logic is the opposite definitions.

The variables perform the same role as in normal algebra, which is to represent a quantity which may take the value of any allowed constant. Boolean variables are usually represented by upper-case Roman characters, whereas normal variables are upper/lower case Roman, Greek, Cyrillic or invented characters. This is not a rule, but a guideline worth following.

Operations are instructions for the reader to perform, and are used in the assignment of values to variables.

e.g. $x = 4y + 1$

means: Take for the value of x the value of 4 multiplied by the value of y plus 1.

In Boolean algebra there are only four operations, and so the whole system is very simple to use. The operations are:

AND	C = A AND B	Take C to be true if both A and B are true. Otherwise take C to be false.
(inclusive) OR	C = A OR B	Take C to be true if either A or B or both A and B are true. Otherwise take C to be false.
(exclusive) OR	C = A XOR B	Take C to be true is either A or B are true. If A = B take C to be false.
NOT	C = NOT A	Take C to be true if A is false. Otherwise take C to be false.

These have shorthands which are in universal use:

AND $\quad$ C = A.B or C = AB

OR $\quad\quad$ C = A + B

XOR $\quad$ C = A $\oplus$ B

NOT $\quad$ C = $\overline{A}$

Also common to both conventional algebra and Boolean algebra is the use of parentheses, braces and brackets - [{ ( ) }] - to signify order of calculation. You perform the operations inside the parentheses before those outside them.

There are many conventional operations which can be performed in different orders, but which return the same result whatever the order. This is expressed as an *identity*, such as those below:

$$x + 5 \equiv 5 + x \quad \text{or} \quad \sin \sin^{-1} x \equiv \sin^{-1} \sin x$$

There are also many Boolean identities, which are listed in the following section.

4.1.3 Boolean Identities

None of these include the exclusive-OR operation. This is because the XOR operation can be expressed entirely in terms of the simpler AND, OR, and NOT operations. What is more, the expression is not unique!

$$AB \equiv BA$$

$$ABC \equiv A(BC) \equiv (AB)C \equiv (AC)B$$

$$AA \equiv A$$

$$A.1 \equiv A$$

$$A.0 \equiv 0$$

$$A(B + C) \equiv AB + AC$$

$$A + AB \equiv A$$

$$A + BC \equiv (A + B)(A + C)$$

$$A + B \equiv B + A$$

$$A + B + C \equiv (A + B) + C \equiv A + (B + C) \equiv (A + C) + B$$

$$A + A \equiv A$$

$$A + 1 \equiv 1$$

$$A + 0 \equiv A$$

162

$$\overline{1} \equiv 0$$

$$\overline{0} \equiv 1$$

$$A + \overline{A} \equiv 1$$

$$A\overline{A} \equiv 0$$

$$\overline{\overline{A}} \equiv A$$

$$A + \overline{A}B \equiv A + B$$

$$\overline{A + B} \equiv \overline{A}.\,\overline{B}$$

$$\overline{A.\,B} \equiv \overline{A} + \overline{B}$$

These last two are *DeMorgan's theorems*.

The XOR identities:

$$A \oplus B \equiv A\overline{B} + \overline{A}B \equiv (A + B)\,\overline{(A.\,B)} \equiv \overline{AB + \overline{A}.\,\overline{B}} \equiv (A + B)\,(\overline{A} + \overline{B})$$

You should practice using Boolean algebra so that you can simplify logic circuit problems before getting down to designing an actual circuit.

4.1.4 Truth Tables

Because, in Boolean algebra, variables can only take two values it is possible to prove relationships simply by writing down every possible value that every variable can take, computing the results of both sides of the expression, and comparing the results. The tabulated data is called a *truth table* and I shall use one to prove the first of DeMorgan's theorems.

A	B	$\overline{A}$	$\overline{B}$	A + B	$\overline{A.\,B}$	$\overline{A + B}$
0	0	1	1	0	1	1
0	1	1	0	1	0	0
1	0	0	1	1	0	0
1	1	0	0	1	0	0

Table 4.1

The last two columns are quite evidently identical. If there are N variables you have to consider 2^N combinations, so the method becomes clumsy for $N \geq 5$.

A truth table is not just a computational device for evaluation of problems in Boolean algebra. They are often used to express the operations that a particular digital circuit performs, or is desired to perform. The designer must then reduce the table into a set of Boolean equations which are equivalent to the table. From these equations a circuit can be designed, by replacing Boolean operations with gates.

4.1.5 Canonical Form of Logic Expressions

All the possible combinations of states of A and B can be represented by a binary number, represented AB, where:

$$AB = A \times 2^1 + B \times 2^0 \qquad \text{(here + = plus)}$$

i.e. $\quad$ A = 0, B = 0 $\quad$ is $\quad 0_2 + 0_2 = 0_2 = 0_{10}$

$\qquad$ A = 1, B = 0 $\qquad 10_2 + 0_2 = 10_2 = 2_{10}$ $\quad$ etc.

So, the truth table (a) could be written as (b). In table (b) AB is written in binary & decimal.

	(a)				(b)	
A	B	F		AB		F
0	0	1		00 0		1
0	1	0		01 1		0
1	0	1		10 2		1
1	1	1		11 3		1

Table 4.2

Looking at table (b) we see that F = 1 when AB = 00 or AB = 10 or AB = 11. In the canonical form this is written as the sum of the decimal numbers representing the states AB.

$$F = \Sigma \ (0,2,3)$$

This form can be useful in the deduction of the Boolean logic expression which represents a truth table. Once you have the expression you can replace each Boolean operation with a logic gate, and you have a circuit which makes the decision represented by the table.

e.g. $\qquad\qquad\qquad F = \Sigma \ (0,2,3)$

$$F = \overline{A}.\overline{B} + A.\overline{B} + A.B$$

$$F = \overline{A}.\overline{B} + A.(\overline{B} + B)$$

164

$$F = \overline{A}.\overline{B} + A$$

$$F = \overline{A + B} + A$$

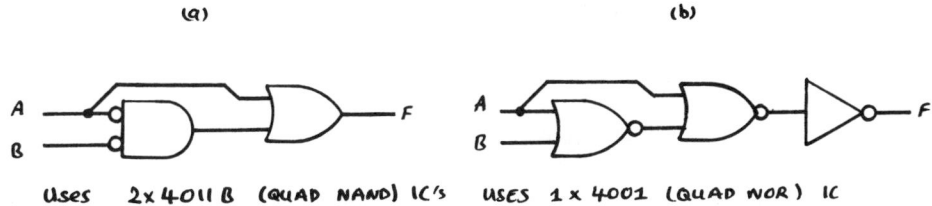

Uses   2x 4011 B (QUAD NAND) IC's   USES 1 x 4001 (QUAD NOR) IC

Figure 4.4 Final circuits as the result of analysis given in text

Figure 4.4 shows, symbolically, the circuit which evaluates F. In actual fact, I would use the alternative circuit (b), because it requires the use of only one IC, not two. have been easier to note that F = AB rather than use the procedure outlined above.)

4.1.6 Karnaugh Mapping

This is a graphical method of logic simplification. The process involves special diagrams, called *Karnaugh Maps*, on which every state of the variables (inputs) is represented. The maps are structured to allow easy logic simplification. You should be able to pick up the structure from the examples below:

Two Variable Map

This represents the general expression F = f(A,B) where f is a *Boolean function*.

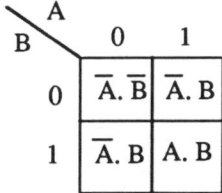

Three Variable Map F = f(A,B,C)

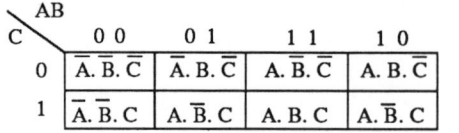 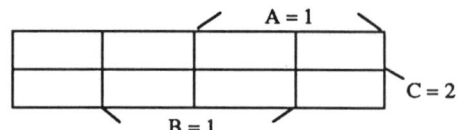

C \ AB	0 0	0 1	1 1	1 0
0	$\bar{A}.\bar{B}.\bar{C}$	$\bar{A}.B.\bar{C}$	$A.B.\bar{C}$	$A.\bar{B}.\bar{C}$
1	$\bar{A}.\bar{B}.C$	$\bar{A}.B.C$	$A.B.C$	$A.\bar{B}.C$

and so on...

Each cell represents a possible combination of the variables, and the left hand side maps have the expression which is true written in the corresponding cell. In order to represent systems with 3 or more variables in 2 dimensions the variables are grouped to represent binary numbers, as in §4.15. With grouped variables only one bit of the group changes as you move from right to left along the rows. With a 4-variable map this would be true along the columns as well. This is so that there are areas of the map in which one particular variable is true.

Procedure

Make a truth table of the required function.

ABC	F
000	1
001	0
010	0
011	1
100	1
101	0
110	0
111	1

Table 4.3

Then fill in a Karnaugh map with the values from the truth table.

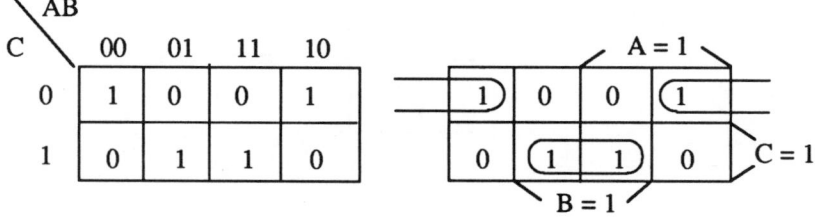

C \ AB	00	01	11	10
0	1	0	0	1
1	0	1	1	0

The special form allows a great deal of simplification to be done by inspection. This is done by *coupling* groups of adjacent cells, as shown above. The couples must always be rectangles containing 2, 4, 8... cells. The edges can be considered to be adjacent. Then the expressions which describe the couples are written down.

The *edge couple* includes two cells in which both B = 0 and C = 0, so I may immediately write down a simple expression to describe it.

$$\overline{B}.\overline{C} = 1$$

The other couple includes two cells in which B = 1 and C = 1. It is also described by a simple expression:

$$B.C = 1$$

$$\therefore \quad F = \overline{B}.\overline{C} + B.C$$

In a simple case such as this you should be able to write down the expression immediately (after a little practice with maps!)

The reason why Karnaugh mapping is so successful is because it exploits the human brain's ability to rapidly recognise visual patterns. Mapping is a much more *natural* method of logic simplification than algebra.

Note on Digital Logic Integrated Circuits

The following sections include details of some of the many things which can be done with simple *combinational logic* circuits. Many of the applications are available in integrated circuit form. I shall give details of the CMOS 4000B series of ICs, which are particularly simple to use. I recommend that you connect up some circuits to get a feel of practical digital electronics. The CMOS ICs have power supply pins V_{DD} and V_{SS}. V_{SS} must always be grounded and V_{DD} should be connected to a potential between 3 and 18 V. The circuits use V_{DD} to represent logic 1 and V_{SS} to represent logic 0. Inputs may be applied to the circuit by directly connecting an IC input to the appropriate power supply voltage. The usual method for displaying outputs is a simple LED circuit, which is lit for a logic 1 output. Further details of the CMOS 4000B and TTL 74LS logic *families* are given in a later section.

4.1.7 Decoders

The circuits discussed here are examples of a class of *combinational logic* circuits (i.e. the output is dependent on the instantaneous value of the inputs) which decode an input *word* (grouping of bits) to give an output. The output is often used as a control input for other circuits. In many cases, the input words represent numbers.

The Digital Representation of Numbers

In a previous section I have shown how a group of bits can be thought of as a binary number. It should be no real surprise that we actually use groups of bits to represent an actual binary number.

These binary numbers are referred to as *words*. Usually words are 4, 8, 16, or 32 bits long.

With	4	bits you can represent	16	different numbers (=	2^4)
	8		256		2^8
	16		65 536		2^{16}
	32		4 294 967 296		2^{32}

The actual number that a particular bit pattern represents is usually given in binary (base 2), octal (base 8), decimal or *binary coded decimal* (base 10), or hexadecimal (base 16).

inputs ABCD	number encoded binary	octal	decimal	BCD	hexadecimal
0000	0	0	0	0	0
0001	1	1	1	1	1
0010	10	2	2	2	2
0011	11	3	3	3	3
0100	100	4	4	4	4
0101	101	5	5	5	5
0110	110	6	6	6	6
0111	111	7	7	7	7
1000	1000	10	8	8	8
1001	1001	11	9	9	9
1010	1010	12	10	-	A
1011	1011	13	11	-	B
1100	1100	14	12	-	C
1101	1101	15	13	-	D
1110	1110	16	14	-	E
1111	1111	17	15	-	F

Table   4.3

The table shows how the four parallel inputs are used to represent numbers in different systems. In the binary system the number represented is simply given by the exact states of the bits (i.e. TRUE is 1_2 and FALSE is 0_2). A is called the *most significant bit*, it represents 2^3, and D is called the *least significant bit*, it represents 2^0. In circuits, the order ABCD must be adhered to or the code will become scrambled.

Although binary *is* the system in which the numbers are represented it is a clumsy system to use because numbers can be large and are mostly incomprehensible. (What does 101010001010101011111001 represent?) The other systems are frequently used to refer to binary numbers, because they are much easier to deal with.

Octal is a base eight number system; numbers in octal should be written with a trailing subscript eight (e.g. $14_8 = 12$ in decimal). Many computer languages accept the prefix &O to mean *in octal* (e.g. &O14).

BCD is a system used because it is easier to convert into decimal than straight binary, octal, etc. This is because the codes representing the numbers 10 to 15 are simply ignored. It is an inefficient encoding system.

Hexadecimal is a base sixteen number system. The numbers 10 - 15 are represented by the letters A - F for compactness. The system is in widespread use in computers. You should use a trailing subscript sixteen to mean in *hex.*, but computer languages tend to accept a suffix H or a prefix &H. (e.g. C_{16} = &HC = CH = 12 decimal).

BCD to 7-Segment Decoders

This brings us to the first application of decoder circuits. That is to take a 4-bit binary input and light up LEDs in a *seven segment display* in a pattern which we interpret as a decimal number.

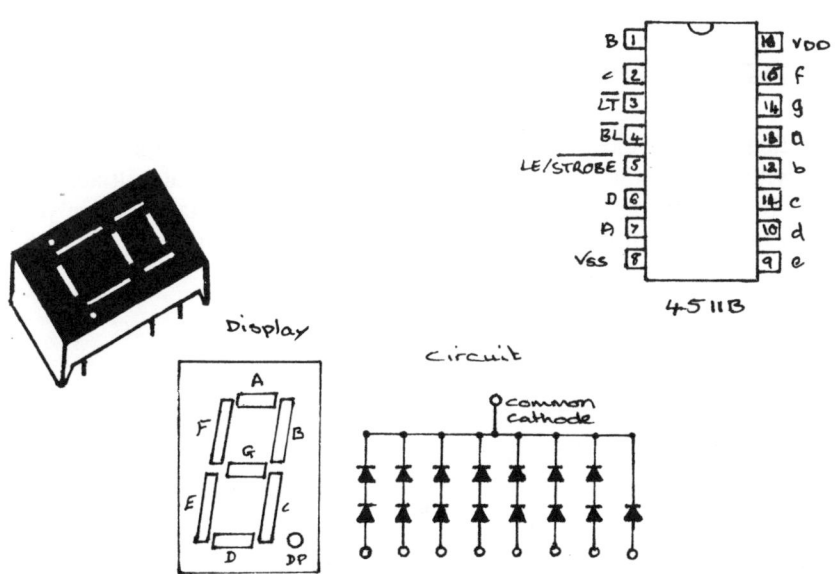

Figure 4.5 BCD to 7-segment decoder circuit

The decoding is done by a complicated logic circuit formed in an IC. The action is best represented by its truth table.

- See Table 4.5 on the next page -

ABC	a	b	c	d	e	f	g	disp
0000	1	1	1	1	1	1	0	0
0001	0	1	1	0	0	0	0	1
0010	1	1	0	1	1	0	1	2
0011	1	1	1	1	0	0	1	3
0100	0	1	1	0	0	1	1	4
0101	1	0	1	1	0	1	1	5
0110	1	0	1	1	1	1	1	6
0111	1	1	1	0	0	0	0	7
1000	1	1	1	1	1	1	1	8
1001	1	1	1	0	0	1	1	9

Table 4.6

A typical IC is the CMOS 4511B. As well as having the four inputs for the binary word (labelled A, B, C, D) it has a *lamp test* input, which forces all LEDs to be on regardless of the input word, and a *blanking* input, which turns all LEDs off regardless of the input word. It has a strobe input which is used to freeze the display. We will see later how this can be done. If the input is not a BCD code the IC displays nothing.

4.2 Arithmetic Circuits

This section takes the electronic representation of numbers, and logic circuits, one stage further with the introduction of arithmetic circuits. Circuits that will add, subtract or multiply by using logic. The most dramatic step you have to take, before you do this, is simply to realise that it is possible. The circuits are not particularly complex.

4.2.1 The Half Adder Circuit

To design a logic circuit, the first stage is usually to write down the desired truth table. I shall do this for a circuit which outputs a value Σ, determined by:

$$\Sigma = A + B$$

To write the truth table, let us examine every possible input and result:

A	0	1	0	1
B +	0 +	0 +	1 +	1 +
Σ	0	1	1	10

Examining the above, we see that we need two bits to represent Σ. It is customary to use S, for sum, and C, for carry.

170

Now we write the truth table...

A	B	C	S
0	0	0	0
0	1	0	1
1	0	0	1
1	1	1	0

Table 4.7

From the table two simple logic equations can be deduced:

$$S = A \oplus B \quad \& \quad C = A.B$$

The symbol on the LHS is a conventional representation for the half adder.

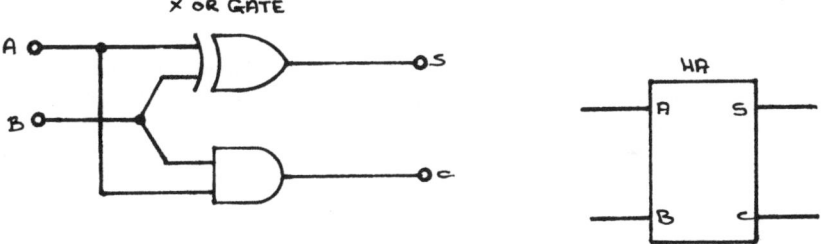

Figure 4.6 1-bit half adder circuit

4.2.2 The 1-bit Full Adder

The half adder generates a carry output, but we really need a carry input so that stages can be cascaded and 2,3,4... bit adders can be constructed.

A *full adder* must perform the action illustrated below:

```
A          1
B          1
CI +       1 +
Σ          11
```

171

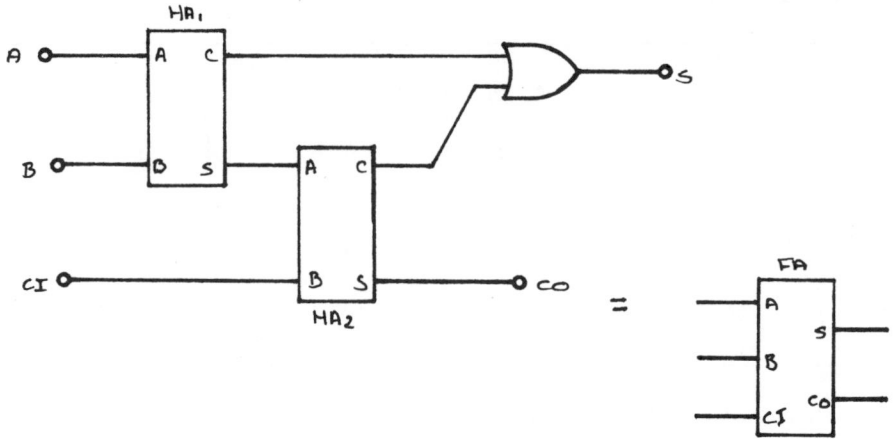

Figure 4.7 1-bit full adder circuit

We represent the sum, Σ, using two bits CO, carry out, and S, sum. The circuit of figure 4.9 does this job. Its operation can be easily deduced from the diagram.

Parallel Adders

Using the 1-bit full adder circuits to add 2, 3, 4... bit numbers can be formed. Such a circuit is called a *parallel adder* if, like fig. 4.2.3, it adds bits of the same order (same power of 2) at the same time.

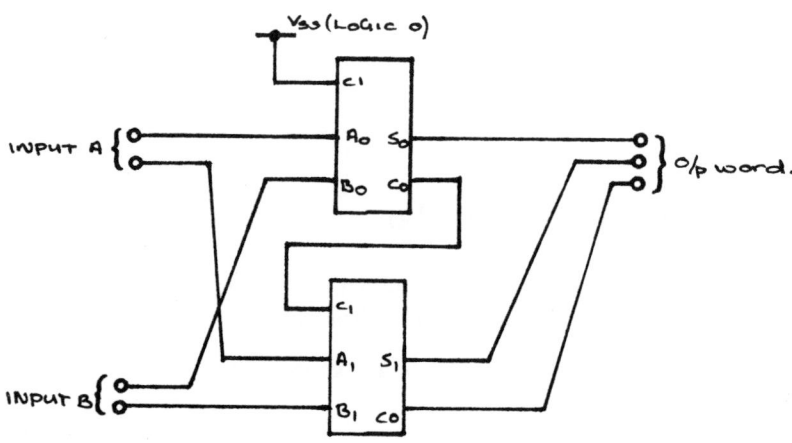

Figure 4.8 2-bit parallel adder circuit

4.2.3 Subtraction

The next logical step is to subtract two numbers. This introduces the problem of the representation of negative numbers by unsigned (assumed positive) binary numbers, which are the only things that logic circuits can deal with.

Sign Magnitude Representation

This solution is the simplest. An extra bit, the *sign bit*, is added to the word. Conventionally, the sign bit is 0 if the number is positive and 1 if it is negative. For example, to represent negative numbers from -7 to +7 four bits are used. The sign bit S and the three magnitude bits ABC. The whole word is usually stored, and written, as SABC. This system is simple to interpret, but it is a wasteful system. In sign magnitude representation both +0 and -0 are possible states of the four bits. The number 0 has no sign, so either choice is meaningless. It is usual to choose +0 to refer to 0, in which case the word 1000 has no meaning and is wasted.

Offset Binary

This coding system is often used in *analogue to digital converters*. It removes any similarity between the states of the logic inputs A, B, C, D and the numbers that they represent. 0000, for example, is chosen to represent the lowest possible number and 1111 the highest possible number.

ABCD	number
1111	+7
1110	6
1101	5
1100	4
1011	3
1010	2
1001	1
1000	0
0111	-1
0110	-2
0101	-3
0100	-4
0011	-5
0010	-6
0001	-7
0000	-8

Table 4.9

The big disadvantage of this system is that it is hard to interpret as a conventional decimal number. Zero is represented by eight (1000), minus one by seven (0111) and plus one by nine (1001). It does not really make much sense unless you walk around with a conversion chart.

The 2's Complement System

This is really the best system. The digits 0 to +7 are represented by the binary numbers 0000 to 0111. The negative numbers are represented by the 2's complement of the positive numbers. They look a little strange, but it is simple to convert to and from the system. Its major advantage is its use in subtraction. To subtract two numbers you simply add the first to the 2's complement (negative) of the second. Just like in real maths! Also the MSB acts as a sign bit as well!

So, how do we form a 2's complement? First, we take the 1's complement of the number, which means we that invert every single bit (i.e. 1010 becomes 0101, 0011 → 1100, etc.). Then we add one. The codes for a four bit number are given in the following table.

ABCD	number	
0111	+7	
0110	6	
0101	5	
0100	4	
0011	3	
0010	2	
0001	1	
0000	0	the 2's complement of 0 is 0
1111	-1	this the 2's complement of +1
1110	-2	etc
1101	-3	
1100	-4	
1011	-5	
1010	-6	
1001	-7	
1000	-8	the 2's complement of -8 is -8

Table 4.10

Sometimes the 2's complement of a number is written *NEG number*, e.g. NEG 5 = -5

Using the 2's Complement to Subtract

To evaluate 4 - 7 in binary, using the 2's complement system, all that is required is the sum 4 + (-7).

The representation of 4 is   0100;
The representation of 7 is   0111;
The 1's complement of 7 is   1000;
∴ the 2's complement of 7 is   1001 (this is the -7 on the table above).

4 plus NEG 7 is: $0100 + 1001 = 1101$

This is a negative number (since the MSB is high). To interpret it as $-1 \times$ a positive number the 2's complement of the result is taken.

$$NEG\ 1101 = 0011\ (3)$$

The calculated answer is, therefore, 4 - 7 = -3, which is correct.

Faults

A fault is introduced by limiting the number of bits available for the magnitude of the number to 3 (i.e. 7 is the maximum magnitude). If you add 0110 (6) to 0101 (5), in a system without signed numbers, you get 1011 (11). But 1011 can represent -5 which is most definitely not 11. You can choose to ignore the negative interpretation, in which case the number correctly represents 11. This *optional interpretation* approach is offered by most computer circuits, allowing 3 bit binary addition and subtraction or 4 bit binary addition to be done with the same circuit.

4.3 Flip-Flops and Sequential Logic

Starting with this section, we enter the world of *sequential logic*. This refers to circuits whose outputs depend on how the states of the inputs have changed through time. For example, with sequential circuits it becomes possible to detect if an input word has represented, in sequence, 4 6 7. This is possible because of circuits, called flip-flops, which are used to *store* a digital number. All sequential circuits involve *pulse generating* circuits of one form or another, so I am beginning this section with a discussion of pulse generators.

All sequential circuit elements can be classified as one of three types of *multivibrator*, which is a circuit whose output changes with time.

Astable multivibrators, or pulse generators/oscillators, are circuits whose output continuously changes from *high* (true) to *low* (false) to high again. The frequency of such changes is usually determined by some sort of RC network. An astable outputs square waves or repetitive pulses.

Monostable multivibrators are circuits whose output is normally in one state, usually low. The output can be forced to go to the other state, i.e. high, but it is not *stable* in this state and *decays* back to the stable state after a preset time. The time is determined by an RC network. The circuits are used to output single pulses of a known length.

Bistable multivibrators, or latches/registers/flip-flops, are circuits whose output can normally be either high or low (i.e. it is stable in both states). The output can be forced to change state. Unlike the monostable it will not decay back to the original state, but will remain in that state until forced back to the original state.

4.3.1 Pulses, Pulse Trains and Square Waves

Before embarking on a discussion of multivibrators it is best to understand a little of the terminology of pulses.

A Pulse

A pulse is generally a signal which changes from its normal state (high/low) to the complementary state (low/high) for a time, and then returns to its normal state.

In an ideal pulse, the states change occurs in zero time, and the states attained are the voltages representing the Boolean constants 0 & 1. In reality, this will never exactly occur. All commercial logic circuits have *noise margins* in their input stages to compensate for this defect, which means that instead of reaching, for example, 5V to represent 1 the input has to lie between 4V & 5V, and to represent 0 it has to lie between 0V and 1V.

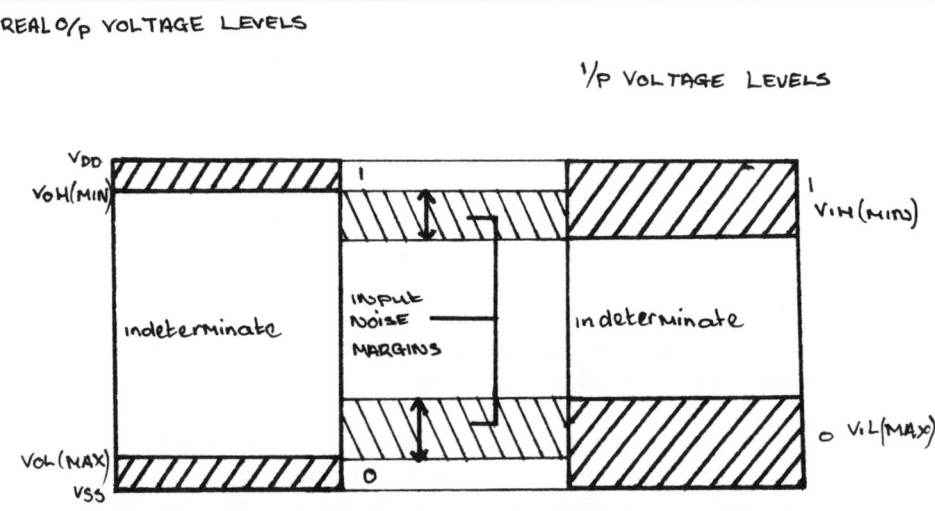

Figure 4.9 Input and output 'characteristics' of logic circuits, showing noise margins

A real pulse also takes time to change state. These faults are represented in figure 4.19. The *amplitude* is, as usual, the difference between the upper limit of the output and the lower limit of the output. The *rise time* and *fall time* of the pulse are defined, respectively, as the time taken for the output to rise from 10% of the amplitude to 90% of the amplitude, and the time taken for the output to fall from 90% to 10% of the output. The *pulse width* (or period) is the interval between the pulse rising to 50% of the amplitude, and falling back to 50% of the amplitude.

A *pulse train* is a repetitive sequence of pulses. The frequency is the reciprocal of the period. A *square wave* is a pulse train with a 50/50 mark/space ratio.

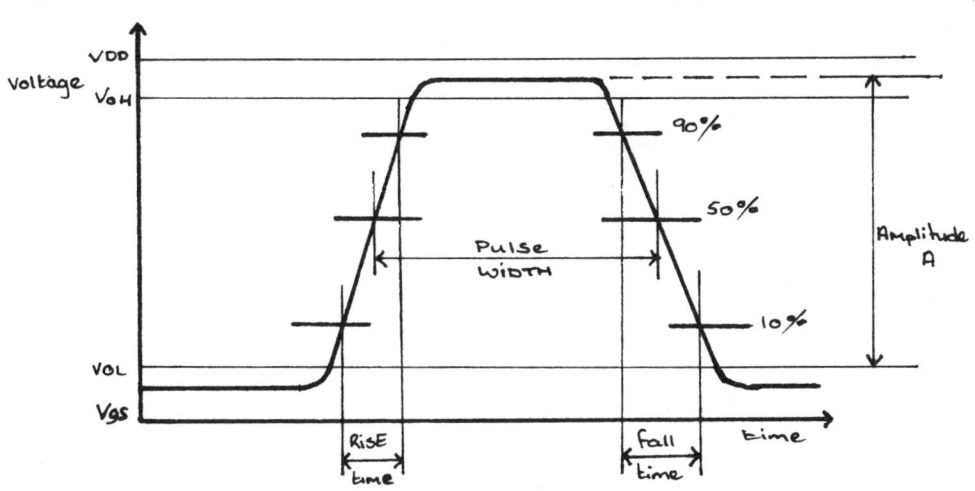

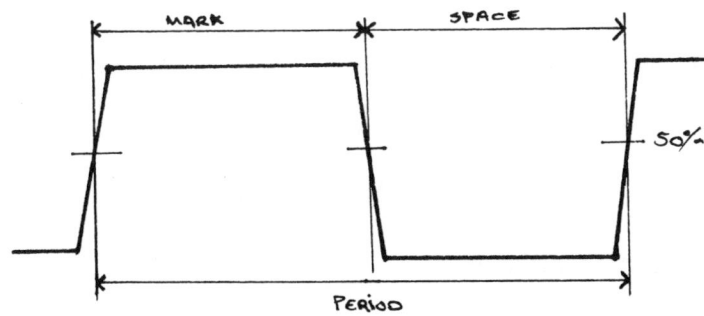

Figure 4.10 Pulse nomenclature

4.3.2 Astables

Schmitt Trigger Relaxation Oscillator (Square Wave)

Perhaps the most basic astable is that formed from a Schmitt trigger inverter. The operation is simple.

Suppose that the capacitor is uncharged. The output is high and the capacitor charges to the upper trigger level. This changes the output to low and so it discharges to the lower trigger level. Again, the output changes and the process repeats.

It is important to remember that connecting an output load robs capacitor charging current and lowers the free running frequency of the circuit. In all astables and monostables formed using gates you should *buffer* the output by connecting it to a follower. A CMOS inverter is ideal because of the very high input impedance of MOSFET circuitry.

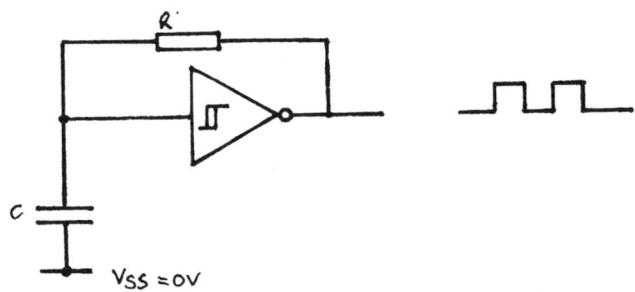

Figure 4.11 Schmidt trigger astable

In such a condition the *free-running frequency* is determined by the hysteresis of the Schmitt trigger, and can be found by following the method used for the UJT oscillator. V_{LTL} and V_{UTL} are the upper and lower threshold levels.

$$f = 1/2RC \ln \{(V_{DD} - V_{LTL})/(V_{DD} - V_{UTL})\} \qquad (4.1)$$

Two Inverter Oscillator

Another, simple, oscillator, in more widespread use than the Schmitt circuit, uses two inverters (plus an extra as a buffer).

If a capacitor is uncharged, and the output of the second inverter is high, there is V_{DD} across the resistor, which causes the capacitor to charge until the voltage across it is sufficient to change the output states - i.e. until it reaches V_{DD}. There is now $-V_{DD}$ across the resistor and so the capacitor discharges to V_{SS}. This process repeats, giving a square wave output. Such an oscillator is seldom used to precision, and detailed calculation of the frequency is not required. It will be approximately 1/2RC.

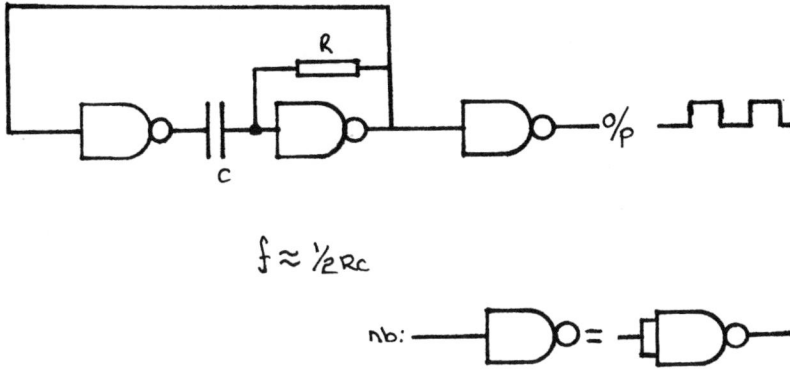

$$f \approx \frac{1}{2RC}$$

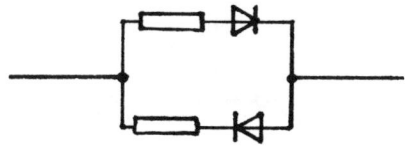

Figure 4.12 Two-inverter astable with buffer

Adjusting the Mark/Space Ratio

The mark/space ratio, sometimes called the *duty cycle*, of a square wave oscillator can be forced away from 50% by making the resistor have different values in each direction.

Figure 4.13 Making a resistor have a different value in each direction

IC Timer Circuits

There is a good range of ICs designed to be timers, which make much more accurate oscillators than circuits built from gates.

Crystal Oscillators

Quartz crystals happen to act as though they are an LC resonant circuit, and by cutting the crystal precisely it can be made to resonate at an accurately known frequency. The crystal symbol, with equivalent circuit, is shown in figure 4.14.

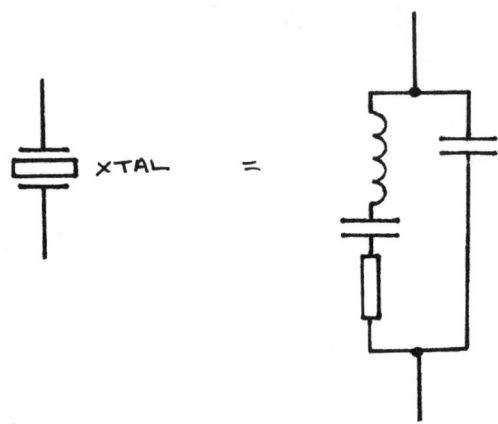

Figure 4.14 Quartz Crystal and equivalent circuit

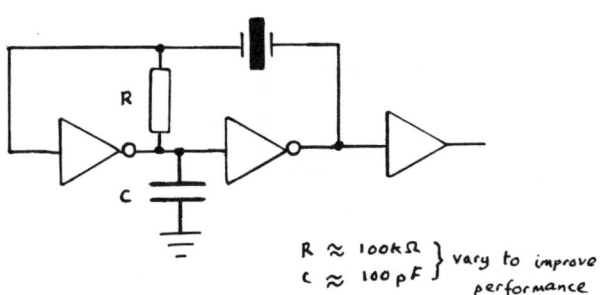

$R \approx 100k\Omega$ } vary to improve
$C \approx 100pF$ } performance

Figure 4.15 Quartz crystal abstable

The circuit shown is a CMOS crystal oscillator. The big advantage of crystal oscillators is that they are very accurate and very stable. However, this is countered by the fact that crystals are only available with resonant frequencies between a few hundred kHz and a few MHz.

4.3.3 Monostables

Simple Monostable

The simple monostable circuit is designed to be triggered from a short positive going pulse. Upon triggering, it outputs a positive going pulse of pre-determined period. This pulse must be longer than that of the trigger pulse.

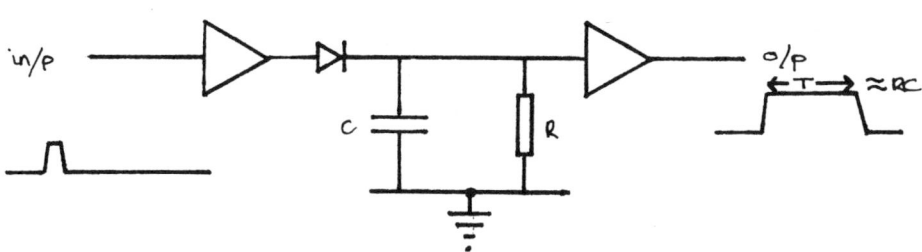

Figure 4.16 Basic (bad) monostable

The circuit illustrates the principle of operation, but will not make a very good practical circuit. When the input is brought high for a short period the capacitor charges to V_{DD} minus one diode drop. As long as the input is high the capacitor stays charged up and so the input to the second gate is high, and the circuit output goes high. When the input returns low the capacitor slowly discharges through R giving rise to a long pulse.

This circuit is not very good because, with a worst case output and a diode drop, the input at the second gate might not reach $V_{IH}(\text{min})$ and so the output pulse will not be generated.

Better Monostable

This two gate circuit is superior, because it will work every time!

If the input is low and the output is low (untriggered) the output of the NOR gate is high and the capacitor does not charge up. This means no p.d. is dropped across the resistor and the input to the inverter is high. So the output is stable in a low state.

When the input goes low the output of the NOR gate goes low and the capacitor starts to charge up. Whilst the p.d. across the capacitor is low enough the circuit output stays high. As soon as the output goes high the circuit is said to be *triggered* and the input can be removed. After an interval, dependent on the time constant RC, the p.d. across the capacitor is sufficient to turn the output off. Both input and output are now off, and the circuit is left in the initial stable condition.

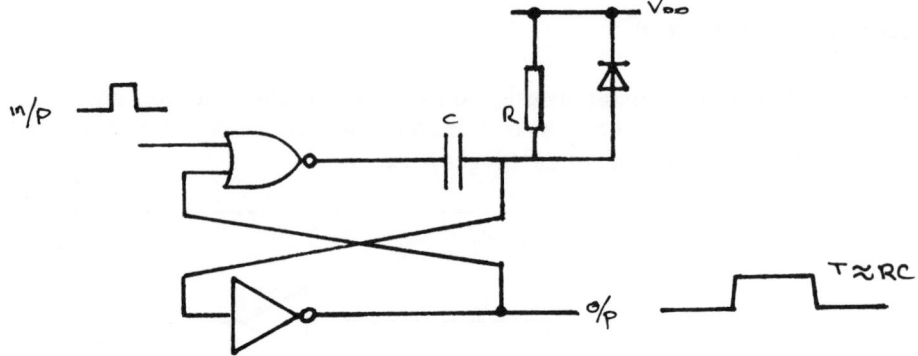

Figure 4.17 Improved monostable

Switch Debouncers

When a switch is depressed it does not cleanly make or break contact, but rapidly makes and breaks contact. This is called contact bounce.

A major use of monostables is to remove this effect. The bouncing lasts for a few milliseconds but this is enough to cause errors in fast counter circuits. A monostable would be used to debounce the switch and eliminate the problem.

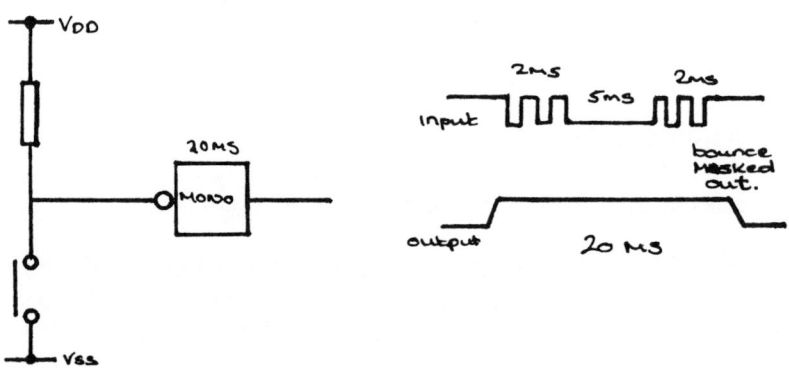

Figure 4.18 Debouncing a switch

4.3.4 Bistables - The SR Flip-Flop or Latch

Circuits Which Are Stable in Both States

There are two basic circuits which are stable in two states, (a) and (b).

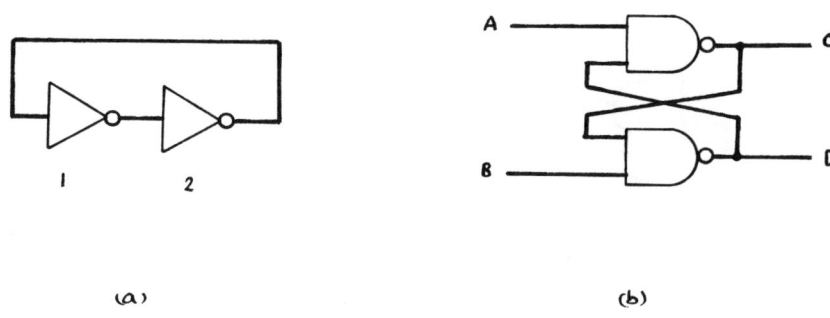

(a)   (b)

Figure 4.19 Bistable circuits

In (a), if the output is high the input to gate 1 is high, and so its output is low. This makes the input to gate 2 low and so its output is high. The feedback forces the gate to stay in the state it is in. If the output is low it is also kept low by feedback, and so it is stable in the low state too.

Circuit (b) is a considerable improvement because it can be forced, by an external stimulus, to change from one stable state to the other. Firstly let us assume that both inputs (A & B) are high.

If output C is high, and output D is low, then the inputs to the upper gate are different and so C is kept high. Also, the inputs to the lower gate are both high so the output is kept low. Therefore, the circuit is stable in this state. Due to symmetry we can immediately say that the circuit will also be stable with C low and D high.

If, when B and D are high and C is low, A is brought low the output of the upper gate becomes 1 which, in turn, forces the output of the lower gate to become 0, and so the circuit enters the other stable state. This means that when the input "A → 0" is applied the circuit changes output from the "C = 0, D = 1" state to the "C = 1, D = 0" state.

The opposite occurs when the input "B → 0" is applied.

With an input "A → 0, B → 0" the output is unpredictable from Boolean algebra. Such an input should not be allowed to occur in real circuits.

The SR Flip-Flop

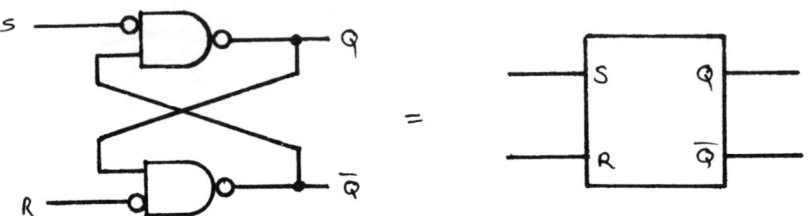

Figure 4.20 The SR flip-flop circuit

The circuit is called an *SR flip-flop*, or latch, and has its own symbol. The outputs are called Q and $\overline{Q}$ (pronounced Q-bar) and the inputs are called S, for set, and R, for reset. If you bring S high you make Q = 1. If you bring R high you make Q = 0.

		before		after		
S	R	Q	$\overline{Q}$	Q	$\overline{Q}$	
0	0	0/1	1/0	0/1	1/0	no change
1	0	X	$\overline{X}$	1	0	x = don't care
0	1	X	$\overline{X}$	0	1	
1	1	unpredictable				disallowed input

Table 4.11

Clocked SR Flip-Flop

This circuit is slightly different. Because of the third input, CLK, the output is only allowed to be changed when the clock (an astable) is high, and so state changes can synchronised.

184

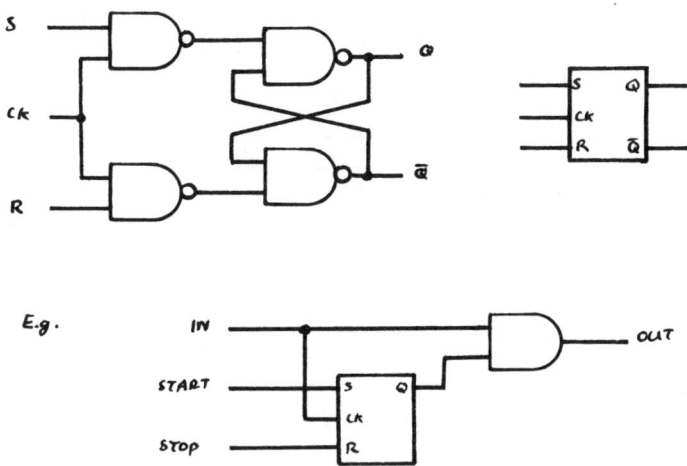

Figure 4.21 A clocked SR flip-flop and application

The example circuit is used to pass a burst of pulses. Such a circuit can be wired with a simple AND gate but the SR-ff version prevents the sequence commencing with half length pulses, which is a common fault of the AND gate version. In this circuit the output state changes on the positive edge of the clock pulse, so this is called a *positive edge triggered flip-flop*. Strictly speaking, this is not an edge triggered flip-flop, since throughout the entire clock pulse the outputs *follow* the inputs, but the clock pulse is usually of shorter period than the *data* pulses.

Logic Races, Negative Edge Triggering and the Master-Slave Circuit

The circuit can produce half length pulses when we try to turn it off. This is best illustrated by examining the circuit waveforms.

There can be a bad pulse transmitted at the end of the sequence. This is called a *logic race*, and is caused by the lack of proper synchronisation and the finite times in which circuit outputs change.

The error can be easily eliminated if a circuit is set up such that the state of the flip-flop changes on the negative edge of the clock pulse. The second set of waveforms illustrate the circuit action when *negative edge triggering* is employed.

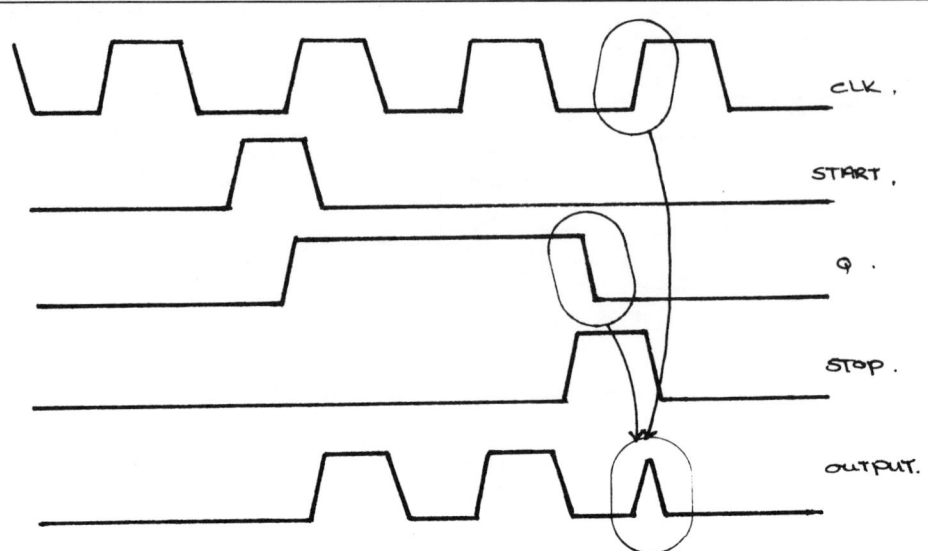

Figure 4.22 Waveforms with positive edge triggering

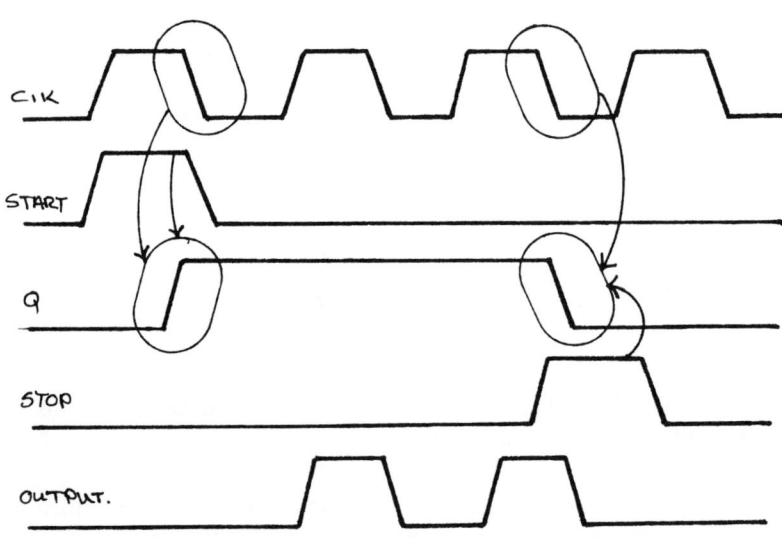

Figure 4.23 Waveforms with negative edge triggering

The negative edge triggering is implemented by using the *master-slave* circuit illustrated. While the clock input is high the Q',Q' outputs, of the master flip-flop, *follow* the S,R inputs but only on the negative edge of the clock pulse is the state of Q',Q' loaded into the slave flip-flop. This action makes sure that the state of the Q,Q outputs is only dependent on the state of the R,S inputs on the negative edge of the clock pulse.

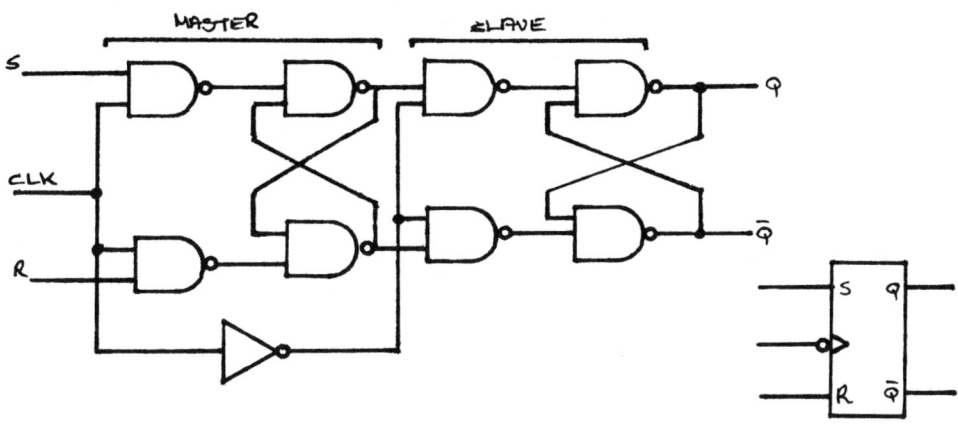

Figure 4.24 Negative edge triggered (master-slave) SR flip-flop

4.3.5 Data Latches and D-type Flip-Flops

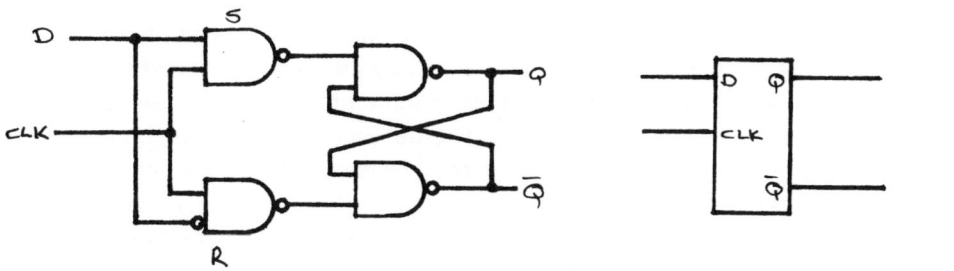

Figure 4.25 The data latch circuit

Instead of using an RS flip-flop as a switch, we often want to exploit the *memory* property of the flip-flop to store a bit. This can be done by making the flip-flop's output, when the clock input is high, change to the state of the S input. In such an application the S pin is usually relabelled D, for *data*. The circuit is referred to as a data latch.

The Negative Edge Triggered D-type Flip-Flop

This circuit is a master-slave data latch, or D-type flip-flop. Because the circuit is edge triggered it stores the state of the D input *at the instant in time* that the clock input state is falling $(1 \rightarrow 0)$.

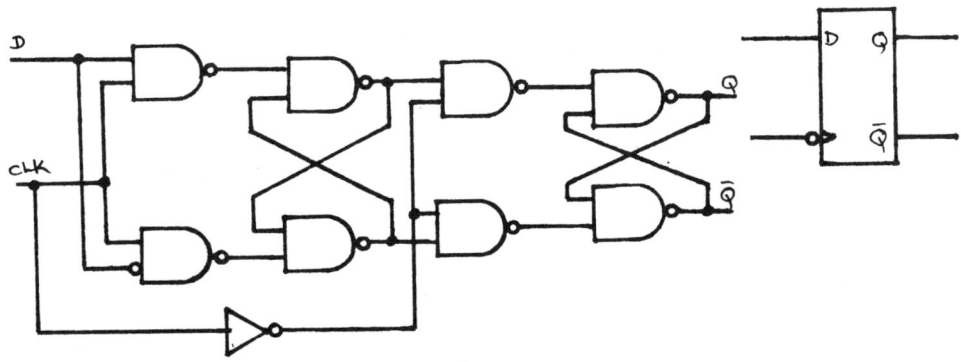

Figure 4.26 Negative edge triggered D-type flip-flop

The Positive Edge Triggered D-type Flip-Flop

In many memory applications it is required that the state of the D input on the positive edge of the clock pulse be stored. This uses a slightly different circuit and because the circuits are not ideal, so the transistors require a finite time to change state, the use of *positive edge triggering* introduces some time-dependent factors which have to be watched. The action of the D-type is characterised by two times: the *set up time* and the *hold time*. The set up time is the time that the data must be steady (i.e. either definitely 1 or 0) before the trigger input goes high, and the hold time is the period for which the data must stay at that level, while CK is high. If both these times are exceeded the data is successfully loaded into the flip-flop. Typical times are nanoseconds.

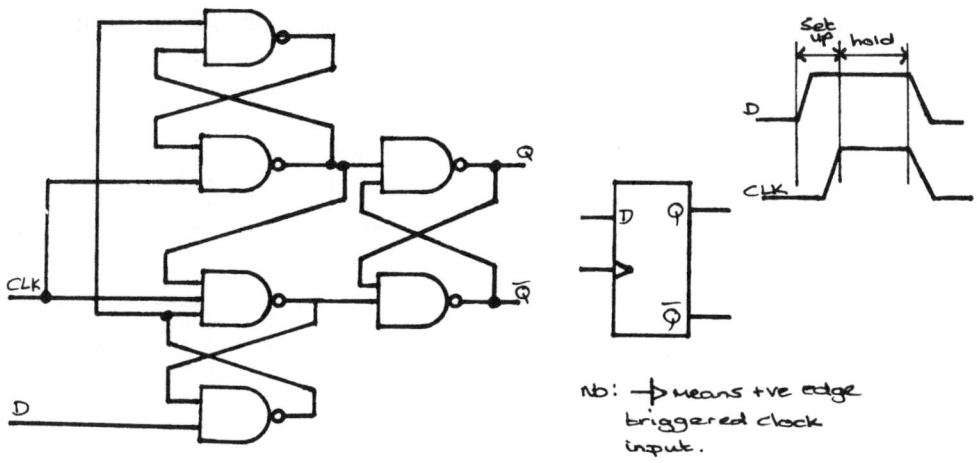

Figure 4.27 Positive edge triggered D-type flip-flop

4.3.6 The T-type Flip-Flop

The T-type

The T-type, or toggle, is a special circuit which can be formed from a D-type. The D input is connected to the $\overline{Q}$ output, which means that Q *changes* state on the edge of the clock pulse.

Because of this *toggle* action the frequency of the output waveform is exactly half that of the input. By cascading T-types you can arrange to divide the input frequency by any power of two.

Counters

A counter is a multi-stage divider in which all the individual Q outputs are used to form the output of the entire circuit. By writing the outputs as a binary number AB, note the order, then the output of the circuit is, in sequence, 00 (0) → 01 (1) → 10 (2) → 11 (3) → 00, etc. This circuit is repeatedly counting from 0 to 3. Such a circuit is called a ripple counter because the effect of the clock pulse ripples through the circuit, as shown in the waveform diagram.

In a ripple counter all the outputs do not change at the same time, and with high *modulo* counters (the modulo is the total number of output states) there can be a

189

significant delay between changes in the first and last flip-flops.

A typical ripple counter IC is the 4024B. It has seven stages, and so can count from zero to 127.

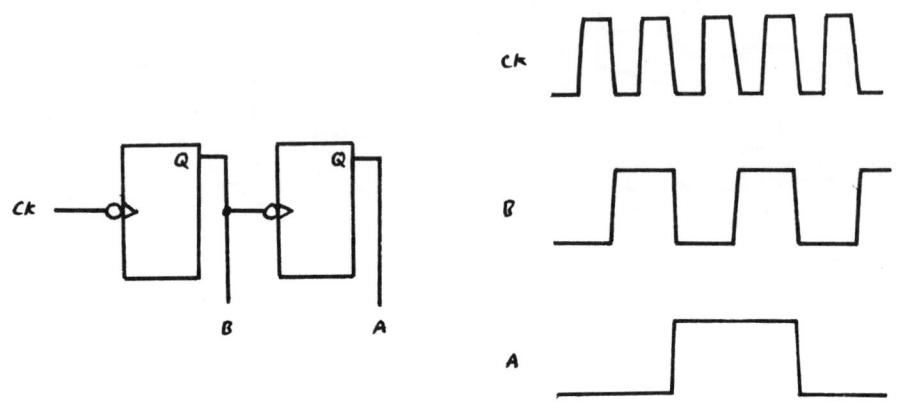

Figure 4.28 Counting with T-types

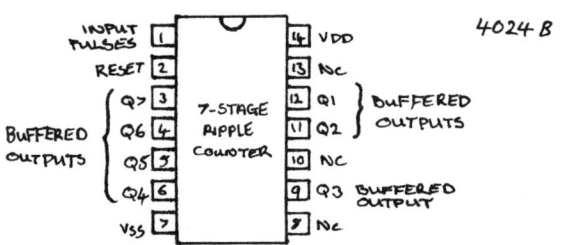

Figure 4.29 The 4027B Ripple Counter

4.3.7 The JK Flip-Flop - An All Purpose Flip-Flop

In the SR-types there is a disallowed state where both S & R are high. It is disallowed because it causes an output condition which makes the logic equations describing the action of the circuit inconsistent, and so circuit action is unpredictable.
The JK-type is a clocked SR-type with an input stage added to prevent the disallowed condition being presented to the output stage. Because of the configuration of the input stage, the output toggles when both inputs are high.

The JK can act as a SR-type and as a T-type.

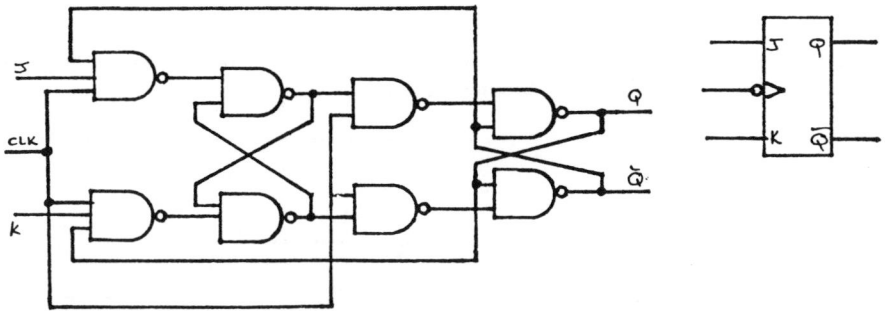

Figure 4.30 The JK master-slave flip-flop

		before		after		
J	K	Q	$\overline{Q}$	Q	$\overline{Q}$	
0	0	0/1	1/0	0/1	1/0	no change
1	0	X	$\overline{X}$	1	0	set
0	1	X	$\overline{X}$	0	1	reset
1	1	0/1	1/0	1/0	0/1	toggle

Table 4.12 Truth Table of JK Flip-flop

An external inverter allows the circuit to be used as a D-type, but more elegant is the provision of multiple AND gated inputs. *Jam inputs* allow the device to be set or reset *synchronously* (on an edge of the clock pulse) or *asynchronously* (at any time regardless of the clock state).

4.3.8 Complicated Flip-Flop Circuits

Synchronous Counters

The problem with ripple counters is that all the outputs do not change at the same time. For example, when a 4-bit ripple counter overflows the output is:

$$\begin{array}{ccc} \text{before} & \text{during} & \text{after} \\ 1111 \rightarrow 0111 \rightarrow 0011 \rightarrow 0001 \rightarrow 0000 \end{array}$$

191

One range of modern CMOS circuits (74AC00 series) can operate at clock frequencies of up to 100 MHz and would detect the outputs $15 \rightarrow 7 \rightarrow 3 \rightarrow 1 \rightarrow 0$, not the desired outputs $15 \rightarrow 0$.

A superior circuit would be one in which the outputs all changed synchronously with an edge of the clock pulse. This can be wired using JK-types.

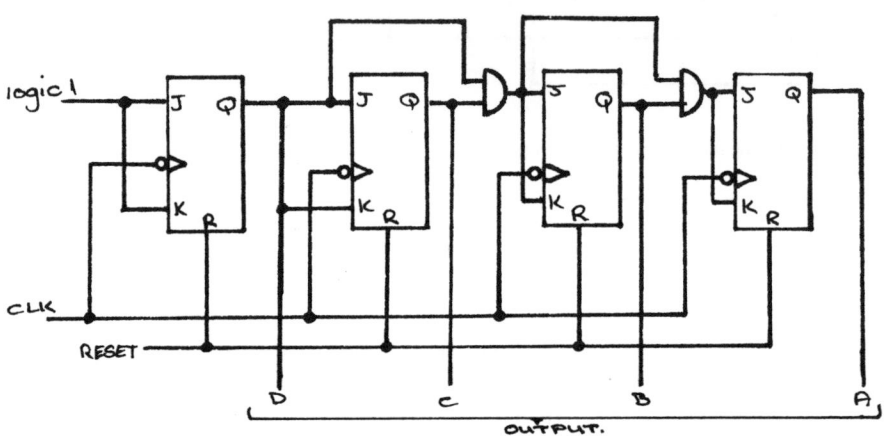

Figure 4.31 Synchronous counter using JK-types

The operation is simple. Each flip-flop is only given *permission* to toggle when the preceding two flip-flops both have the output 1. If this condition is satisfied, it toggles on the clock pulse.

Circuit of 74HC163 Counter

A huge variety of different counters are currently manufactured by semiconductor firms. The complexity of the operation of modern types can be quite sophisticated and, as a consequence, the circuit becomes more and more intricate. As an example I am including the internal circuit diagram of the Texas Instruments 74HC163 counter.

- See Figure 4.32 on the next page -

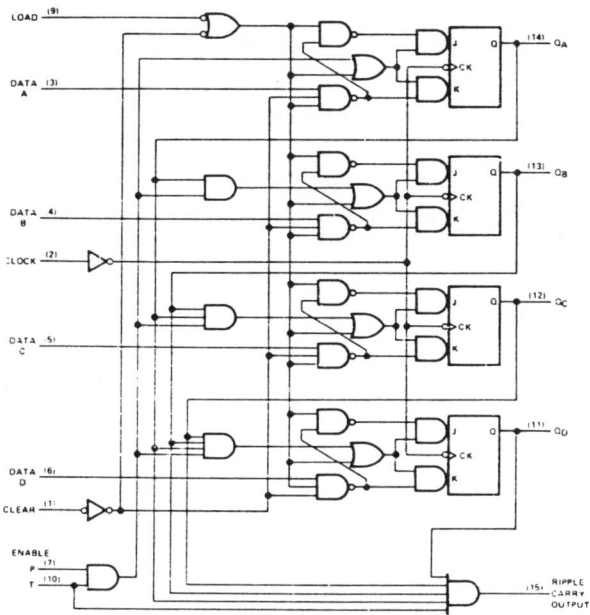

Figure 4.32 Block Diagram of '163 counter IC. Courtesy of Texas Instruments

Shift Registers

If you cascade D-types instead of T-types you get a shift register. A shift register is a *row* of memories. On every clock pulse the data is shifted one place further into the row, until it finally emerges at the output.

The simplest type is the serial in - serial out (SISO).

input	no. clk pulses	output	
X	0	X	
1	1	X	
0	2	X	
1	3	X	
0	4	1	
0	5	0	
0	6	1	
0	7	0	etc.

Table 4.13 A SISO shift register

193

Other types:	serial in - parallel out	(SIPO)	1-bit input	4-bit output
	parallel in - parallel out	(PIPO)	4-bit	4-bit
	parallel in - serial out	(PISO)	4-bit	1-bit

Pseudo-Random Number Generators

A simple circuit can be made from a shift register to generate a sequence of random numbers. It is called a pseudo random sequence because, when the end of the sequence is reached, it just repeats. With a 4-bit SIPO shift register a sequence of fifteen numbers is generated.

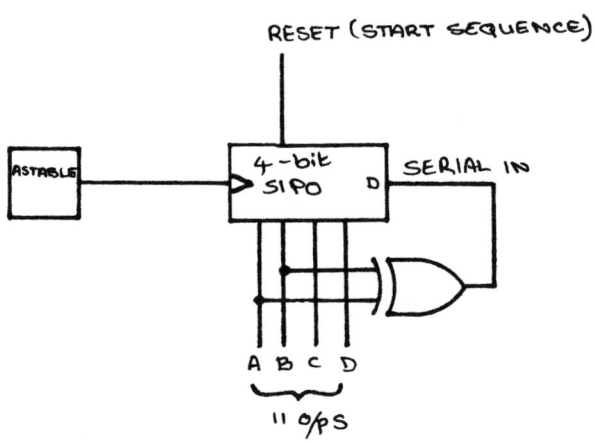

Figure 4.33 A pseudo-random number generator

4.4 The TTL and CMOS Logic Families

One of the major advantages of the widespread manufacture of integrated circuit digital electronics is that the majority of digital work can be done without detailed knowledge of the actual structure of the ICs. However, to appreciate the finer points of gate operation it is necessary to know how they work. This is what I attempt to explain here.

There are three basic sorts of digital IC. A very large family of devices are produced using MOSFETs. These include the CMOS 4000B series, which I have been using in this chapter, and the newer 'silicon gate' CMOS which are essentially very high quality versions of the 4000B circuits. Also, there are two main types of circuit using bipolar transistors: TTL (transistor transistor logic) and ECL (emitter collector logic). TTL is in almost universal use for general high speed applications and most microprocessors, even though they are fabricated in NMOS or PMOS, have TTL compatible outputs. ECL are

194

very high speed circuits and are not really in general use. Most mail order suppliers stock most types of CMOS and TTL, but do not stock ECL at all.

4.4.1 Hardwired Logic Circuits: DRL, RTL, DTL

Diode Resistor Logic

The simplest logic gate can be made from diodes and resistors. The basic gates are the OR gate and the AND gate.

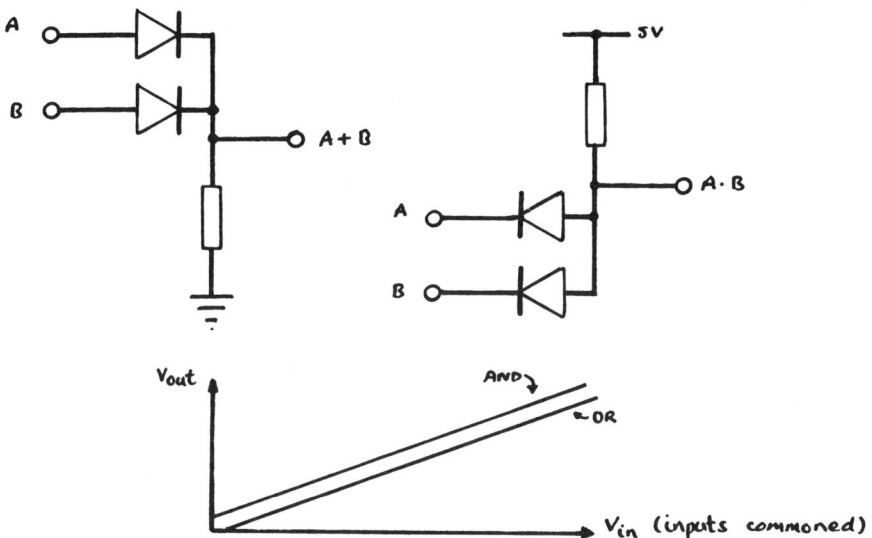

Figure 4.34 DRL AND and OR gates

The operation of the gates is simple, and the characteristics are reproduced in figure 4.34. DRL has a number of major faults. These include:

badly defined high and low levels;

0 dB current gain;

cascading gates can affect the operation of earlier gates; no NOT gate can be formed;

input impedance could be higher; and

output impedance could be lower.

Resistor Transistor Logic

NOT gates can be formed with resistors and transistors, see Figure 4.3.

This gate is much better because it has well defined high and low levels, with a rapid transition between the levels and current gain. A NOR gate can also be formed by adding an additional input.

Diode Transistor Logic

By combining DRL and RTL into DTL, fairly good gates can be formed. This has acceptable current gain and the gate has a good fan out (this is the number of inputs which can be driven from one output with all the cascaded gates kept out of the transition region). See Figures 4.1 and 4.2.

4.4.2 Complementary MOSFET Logic (CMOS)

Of course we do not have to use bipolar transistors at all. MOSFETs are an alternative which give an incredibly high input impedance, an acceptably low output impedance, and a fast transition from high to low.

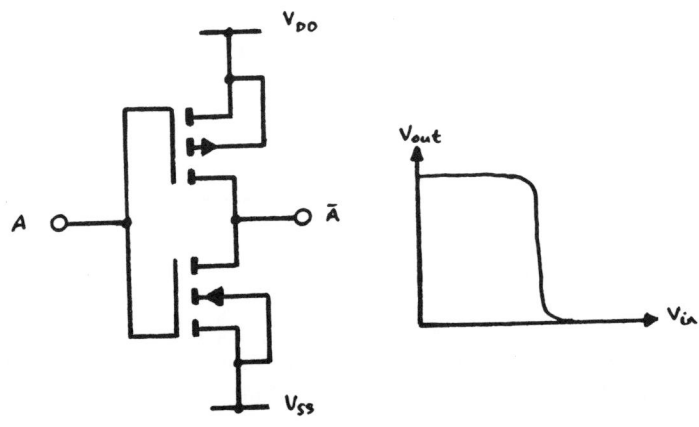

Figure 4.35 CMOS NOT gate

MOSFETs can also be made very much smaller than bipolar transistors, so CMOS is ideal for the production of integrated circuits. The CMOS logic family is built around two basic gates. The inverter (fig. 4.35) and the transmission gate, which I have discussed earlier.

All the basic gates (NOT, NOR, NAND) can be formed from a few CMOS elements, and the XOR/XNOR gates can be built from simpler circuits than their simplest *gate equivalent circuit.*

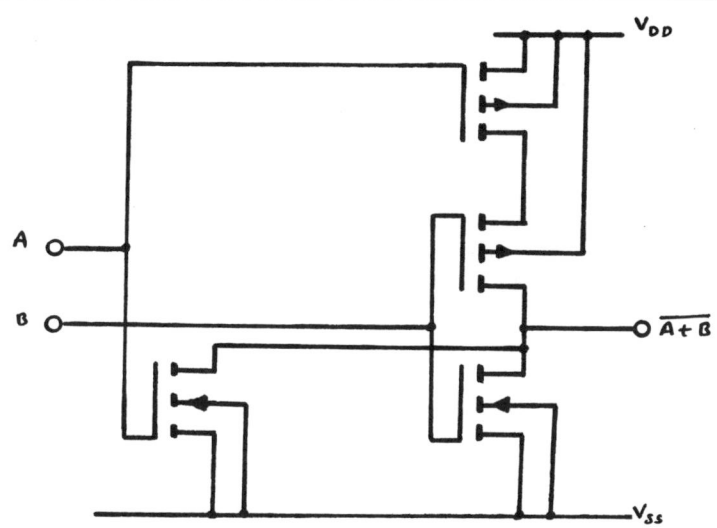

Figure 4.36 CMOS NOR gate

Other advantages afforded by the use of CMOS logic is the almost insignificant quiescent power consumption, making them ideal in remote control type applications, wide range of power supply voltages and reasonably high switching speed.

Full CMOS Specification (4000B series)

parameter	symbol	min	typ	max	@VDD	unit
supply voltage	V_{DD}	3	3 to 15	18	V	
high level input voltage	V_{IH}	3.5		5	5	V
		7		10	10	
		11		15	15	
low level input voltage	V_{IL}	0		1.5	5	V
		0		3	10	
		0		4	15	
high level input current	I_{IH}		10 p	300 n	15	A
low level input current	I_{IL}		10 p	300 n	15	A

Parameter	Symbol						Units
high level output voltage	V_{OH}	4.95				5	V
		9.95			10		
		14.95			15		
low level output voltage	V_{OL}				50		mV
high level output current	I_{OH}	-0.16	-1			5	mA
		-0.4	-2.6			10	
		-1.2	-2.6			15	
low level output current	I_{OL}	0.44	1			5	mA
		1.1	2.6			10	
3		6.8				15	
output 'off' current (3-state)	I_{OZ}				±1.6	15	µA
propagation delay (4011B)	t_{PLH}		125	250		5	ns
			50	100		10	
			40	80		15	
power dissipation quiescent per gate (4011B)	P_D				0.6		µW
clock frequency (counter)	f_{max}		5				MHz

Table 4.15

In these tables: VSS = 0V; positive currents flow into the device under test.

Fuller details of test conditions, calculation of parameters, etc., can be found in manufacturers' data books. The 'High-speed CMOS Logic Data Book', published by and available from Texas Instruments, is excellent. Although it discusses the newer 74HC range of devices, not the 4000B series, the designer's information is very useful.

The 74HC series are superior devices to the 4000B series, and I would recommend their use in all new designs. The reasons why I have not recommended their use in this chapter are that: i) the range of supply voltages they can run from is smaller (2 - 6V); ii) the full range might not be available from most suppliers whereas the full range of 4000B devices probably is; and iii) they are more expensive.

4.4.3 Transistor Transistor Logic

Standard TTL

TTL is a logic family in widespread use. The basic circuit is the NAND gate which has a circuit descended from the DTL AND gate.

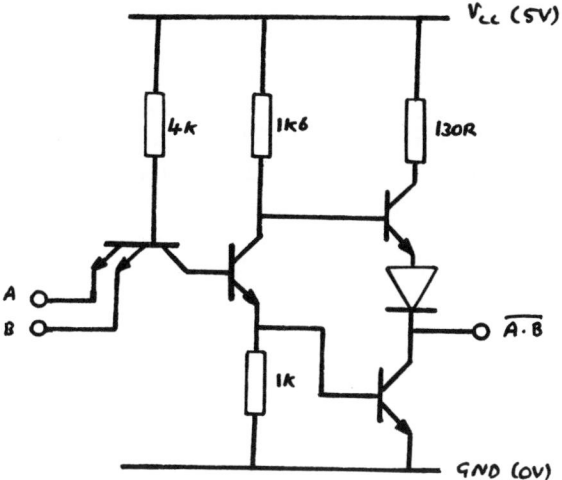

Figure 4.37 Standard TTL NAND gate

The input stage diodes of the DTL gate are incorporated in the multi-emitter input transistor. The multi-emitter transistor, in fact, acts just like the diodes do in the DTL gate. The second stage has a high gain, which gives a small transition region (and, therefore, well defined high and low levels) and an inverting action. The *totem pole* output stage ensures that the gate can source or sink fairly high currents, i.e. it has a low output impedance in both states.

A huge variety of TTL gates are manufactured, and they can be recognised by the 74 or 54 prefix in the serial number. 74 signifies the standard commercial device, and 54 the more robust *military* versions. The military range can be used over a larger range of temperatures, are less sensitive to damage by vibration, etc.

Low Power Schottky TTL

The LSTTL range (prefix 54LS or 74LS) are much more widely used versions of TTL. The circuit has been altered to give lower quiescent power consumption, and Schottky transistors are used to speed up switching.

The Schottky transistor is, in effect, a standard bipolar transistor with a Schottky diode connected across the base-collector junction. When a transistor saturates, the collector is brought to such a low potential (by the external circuit) that the base-collector diode becomes forward biased. The BC current then robs BE current and so the collector current drops, causing the collector potential to rise slightly and reducing the forward bias of the BC diode. This negative feedback effect means that the collector of a

199

transistor never drops below about 100 mV. When V_{CE} reaches 100 mV the transistor is saturated.

Whenever a diode is reverse biased it acts like a capacitor and charge becomes stored on either side of the depletion layer, forward biasing the diode removes the stored charge, but this cannot happen in zero time because of stray resistance which occurs in the junction. This means that the speed with which a transistor can be brought in and out of saturation is limited by an RC network which is part of the base-collector circuit. The Schottky diode, in a Schottky transistor, acts as a clamp forcing V_{BC} to about 400 mV (V_{BE} - V_F) and so saturation cannot occur. The switching speed of a logic circuit built with such transistors is greatly improved.

Full LSTTL Specification

parameter	symbol	min	typ	max	unit
supply voltage	V_{CC}	4.75	5	5.25	V
high level input voltage	V_{IH}	2		5.5	V
low level input voltage	V_{IL}	0	0.8		V
high level input current	I_{IH}			20	μA
low level input current	I_{IL}			-0.4	mA
high level output voltage	V_{OH}	2.7	3.4		V
low level output voltage	V_{OL}		0.25	0.5	V
high level output current	I_{OH}			-400	μA
low level output current	I_{OL}			8	mA
output 'off' current (3-state)	I_{OZ}			± 20	μA
propagation delay (74LS00)	t_{PLH}		9	15	ns
quiescent per gate (74LS00)	P_D	2			mW
clock frequency (counter)	f_{max}	40			MHz

Table 4.16

As you can see, the LSTTL devices are very much faster than CMOS, and have a much higher output current source/sink capability. The price you pay for this is much higher power consumption and much more restricted power supply voltages. The main problem, I have found, that a CMOS designer gets when converting to TTL is that the TTL gate input must source about 0.2 mA for it to admit that you have connected a low input to it. This is because of the structure of the input circuitry. Unconnected CMOS inputs can float at any voltage between V_{DD} and V_{SS}, but an unconnected TTL input generally behaves as if it were connected to logic 1. It is not good practice to leave unconnected inputs, and for TTL unused inputs should be connected to VCC via a 1k resistor. Up to 25 inputs can be connected to each resistor.

The TTL totem-pole output can cause voltage spikes on the V_{CC} and GND lines, since it is a near short between V_{CC} and GND when it is changing state. To eliminate this effect, which can cause stray transitions, connect a 0.1 μF capacitor across the power supply pins of every other IC. Each buffer, counter, or 'big' IC should have its own decoupling capacitor.

4.4.4 Interfacing CMOS to TTL and TTL to CMOS

CMOS to TTL

The big problems with all interfacing are mismatched logic levels and insufficient output current capabilities. The CMOS to TTL interfacing is quite straightforward because of the good high and low levels provided by MOSFET circuitry. The output current capability, however, is sometimes insufficient to make TTL recognise an input as low. This is countered by connecting a CMOS buffer, run from the TTL power supply, to the TTL input. The good CMOS logic levels mean that a circuit run from a supply higher than 5 V will cause proper transitions at the input of a CMOS gate run from 5 V. The CMOS inputs have protection diodes, so a 9 V CMOS to 5 V CMOS connection will not damage the gates, but extra care should be taken if high voltage CMOS is connected to 5 V CMOS. Low voltage CMOS, because of the low TTL $V_{IH(min)}$ of 2 V, would probably switch TTL correctly, and so the buffer can take it's power from the CMOS supply instead of the TTL supply.

TTL to CMOS

Mismatched logic levels are much more of a problem when TTL is driving CMOS, the TTL $V_{OH(min)}$ often being too low to appear as a high input to the CMOS gate. The input voltages are lifted by using a pull-up resistor.

The value of the pull up resistor can be calculated, but accuracy is not very important. A value of about 4k7 is suitable for most cases; if many inputs are being driven from the TTL output, and the CMOS gates are not making proper transitions, you should decrease it by a few kΩ; if the CMOS gates are powered by a 10-18 V supply you should increase it by a few kΩ to ensure the TTL output does not sink too high a current.

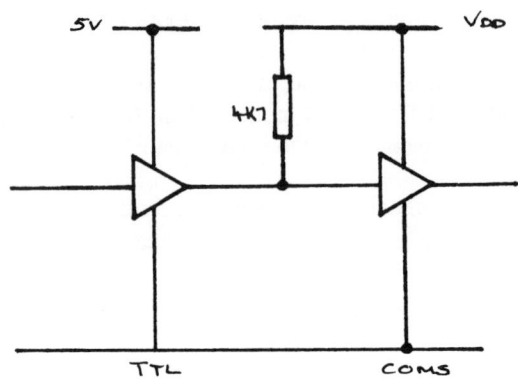

Figure 4.38 Interfacing TTL to CMOS

Questions for Chapter 4

4.1   Design a circuit to indicate which of the following statements is true:

i   The input voltage is less than one volt;
ii   The input voltage is greater than four volts;
iii   The input voltage is between one and four volts.

        Such a circuit is known as a logic probe and is used to determine if a particular terminal is at a good logic level or 'floating'. It is a useful laboratory instrument.

4.2   Show that the circuit of figure 4.i is a 2-input exclusive-OR gate.

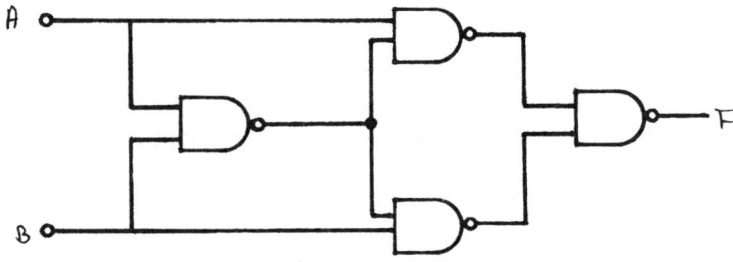

Figure 4.i A logic circuit

4.3   Show by truth table that the Boolean XOR identities, of §4.1.3, are equivalent.

4.4   Prove the following Boolean identities by algebra:

i $A + AC = A$;
ii $(A + B)(A + C) = A + BC$;
iii $A + AB = A + B$.

        hint: start by considering $A(B + C)$

4.5   Reduce the following expressions:

i $A\overline{B}C + AB\overline{C} + ABC$
ii $ABC + \overline{A}\overline{B}C + AB\overline{C} + ABC$

4.6   Design circuits to produce a TRUE output if:

i   The binary input is 5, 6 or 7.
ii   The binary input is 1, 2 or 6.

4.7   Using only five NAND gates, design a half-adder circuit.

4.8   Change the circuit of question 4.7 into one which only uses NOR gates.

4.9   Design a circuit to perform two bit binary multiplication.

        i.e. $P_3P_2P_1P_0 = A_1A_0 \times B_1B_0$

4.10   Change the circuit of question 4.9 into one which either uses NAND gates or NOR gates only.

4.11   Explain the advantages of negative edge triggered flip-flops when compared to pulse triggered flip-flops.

4.12   Design a two input CMOS NAND gate using n-channel and p-channel enhancement mode MOSFETS.

5

AUDIO CIRCUITS

5.1 Simple Passive and Active Filters

5.1.1 Passive Filters Revisited

In chapter 1 I showed how simple filters (frequency dependent attenuators) could be built from simple RC and LR networks. These filters are often very useful, being a principal element in hi-fi tone control networks, but do have a number of drawbacks.

A simple RC network, which gives a cut at frequency $1/2\pi RC$ and whose transfer characteristic *rolls off* at -20 dB/decade (one pole) is shown below.

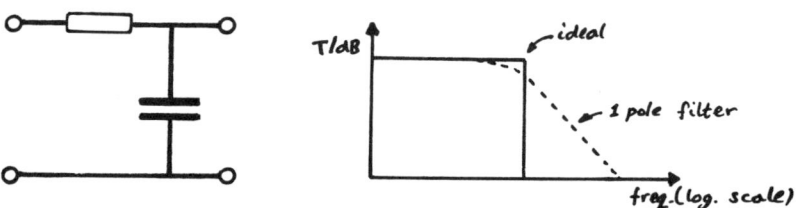

Figure 5.1 A passive filter

This does not compare very favourably with the graph of an ideal filter. We might try to increase the roll off to 2, 3, 4... poles (-40, 60, 80... dB/decade) by adding networks in parallel, but there are drawbacks to this.

The first is that the performance of the filter is altered, since the second network loads the first, and this degrades the action of the filter. The second is that, although this procedure can produce very sharp roll off at high frequencies, the filter will still have the same undesirable *soft shoulder* as the original filter.

The solution to this problem is to include active elements (usually operational amplifiers) in the network. However, this does not mean that we should totally abandon passive filters.

5.1.2 The Tone Control Circuits

A circuit with widespread use is the hi-fi tone control. This circuit must be able to boost, or cut, the high frequency and low frequency components of an audio signal. In such an application a soft shoulder is required.

Bass Cut Circuit

The bass cut circuit has to provide a variable degree of attenuation at low frequencies. The usual practice is to set up a simple RC high pass filter, and alter R so that the breakpoint frequency can be moved. This has the effect of altering the amount of cut at any one specific frequency.

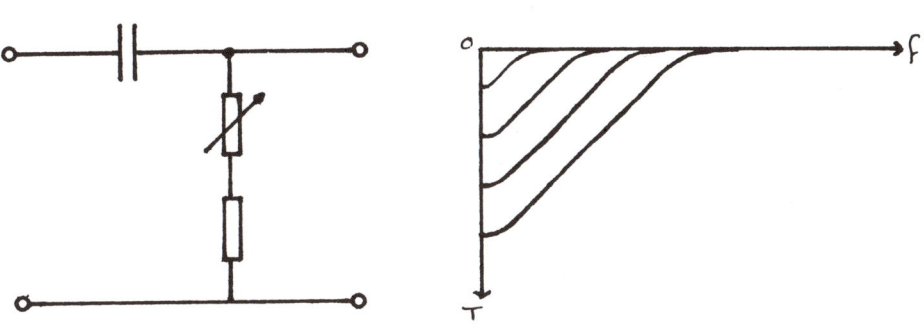

Figure 5.2 Bass cut circuit

The minimum breakpoint of approximately 15 Hz gives, in effect, zero bass cut, since this is below the minimum frequency that most people can hear. The maximum breakpoint is usually put at around 200 Hz.

Bass Boost Circuit

Using a passive circuit, it is impossible to apply boost (gain), but it can be faked by attenuating all frequencies above 200 Hz by a set amount, and then varying the attenuation of lower frequencies between 0 and the set amount. If we add a simple amplifier with a gain equal to the set cut this will, of course, give an actual boost.

The circuit is very simple to understand. Above a critical frequency the capacitor can be assumed to be a short, and so the attenuation takes a value determined by R_1 & R_3. At much lower frequencies the capacitor can be assumed to be an open circuit, and the attenuation is determined by R_1 and $R_2 + R_3$.

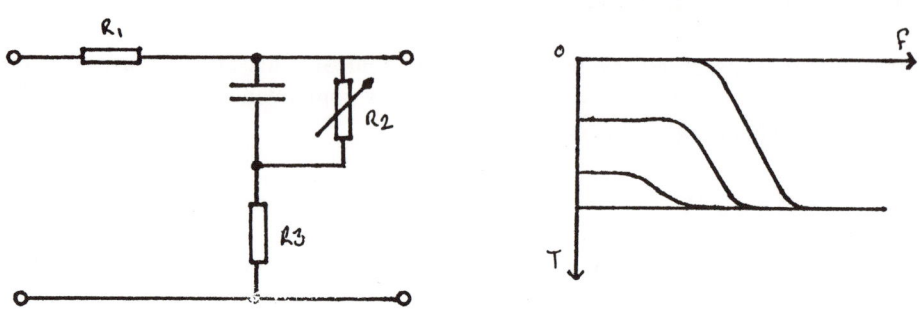

Figure 5.3 Bass 'boost' circuit

Treble Boost and Cut

It is trivial to design circuits along the same lines, to give treble boost and cut.

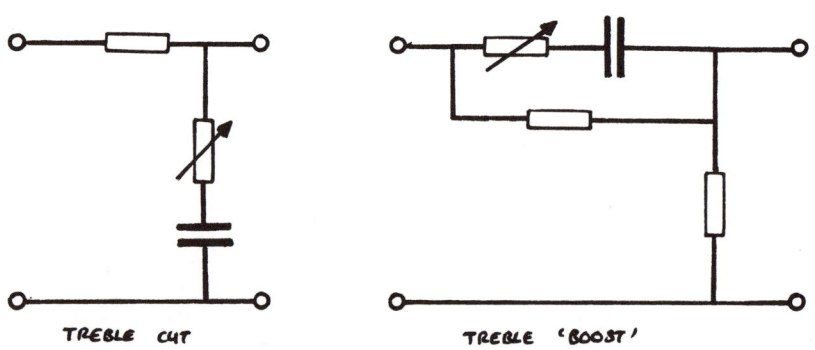

TREBLE CUT   TREBLE 'BOOST'

Figure 5.4 Treble cut and 'boost' circuit

5.1.3 Negative Impedance Converters and Gyrators

Capacitors and resistors are generally easy to work with, and can be easily formed in an integrated circuit. Inductors, on the other hand, are large and heavy, and generally cannot

206

be integrated. This is a problem, since there is often need for an impedance which has the form $j\omega L$.

Negative Impedance Converter

The NIC is a circuit which allows us to turn a capacitor (with impedance $-j/\omega C$) into a component with impedance $j/\omega C$.

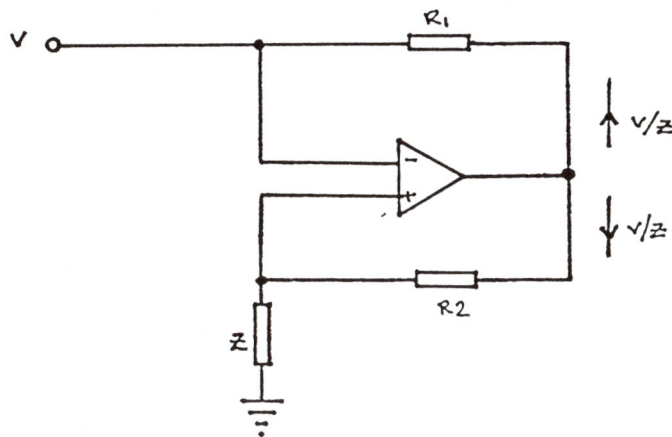

Figure 5.5 Negative impedance converter

$[R_1 = R = R_2]$

If a p.d. V is applied to the input, assuming that the operational amplifier is ideal, the current flowing through the impedance Z will be $I = V/Z$. Since the inputs of an ideal amplifier may not source current this means that it must flow in the same direction through R_2, and so the amplifier output must be at the potential $V + IR$. Therefore, the current flowing through R_1 is $\{V - (V + IR)\}/R$ or $-I$. Evaluating $Z_{in} = V/I_{in}$, gives:

$$Z_{in} = -Z$$

The Gyrator

We are now half-way to the inductorless inductor. The NIC'd capacitor is a reactance with the correct sign, but it has the wrong frequency dependence.

The network of figure 5.6 changes that frequency dependence.

The analysis of the gyrator is also simple. The NIC on the RHS means that the circuit has a series resistance:

$$R + (-R) \ || \ (R + Z)$$

$$= R - R(R + Z)/Z$$

$$= (RZ - R^2 - RZ)/Z$$

$$\therefore \qquad\qquad Z_{in} = R^2 \ /Z$$

Using a capacitor gives: $\qquad Z_{in} = j\omega R^2 C$

which is an inductor of inductance $R^2 C$.

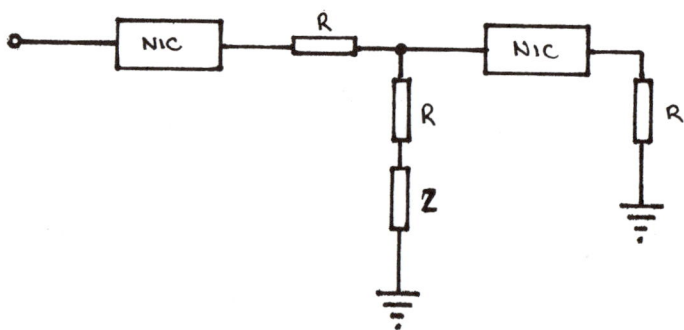

Figure 5.6 Gyrator

A gyrated capacitor can be used in place of most inductors and, therefore, extremely compact filters can be built. There is one fault, but it is not usually a problem. That is that we only have access to one terminal of the inductor; the other has to be tied to ground. It is possible to build active inductors, such as the gyrator, in which both terminals are free, but these often require four operational amplifiers.

5.1.4 Equalisers and Pre-amplifiers

Audio systems are quite complicated in that they require a signal to be changed from a number of different forms into an electrical wave and then, from that, back into a local pressure wave (sound).

Ideally, all these transducers will have a flat response. That is a certain r.m.s. input (be it air pressure as in a microphone, magnetic field in a cassette playback head, or the depth of a record groove) will produce an output of r.m.s amplitude which is dependent on the amplitude of the input, not on the frequency of the input.

We do not live in an ideal world, and every type of transducer has a behaviour which depends on frequency.

Pre-amplifier

To compensate for this effect we must connect the transducer to an amplifier, which is known as a pre-amplifier, that has the opposite frequency response. A simple circuit can usually be built by putting reactive components in the feedback loop of an op-amp.

There are as many types of pre-amp as there are types of transducer. A very common device is an amplifier which equalises the input from a magnetic cartridge (record needle). This must have the *RIAA response curve*.

Transducers in common use are piezo-electric crystals, which usually have a very high output impedance, and magnetic coils, whose impedance is proportional to the frequency.

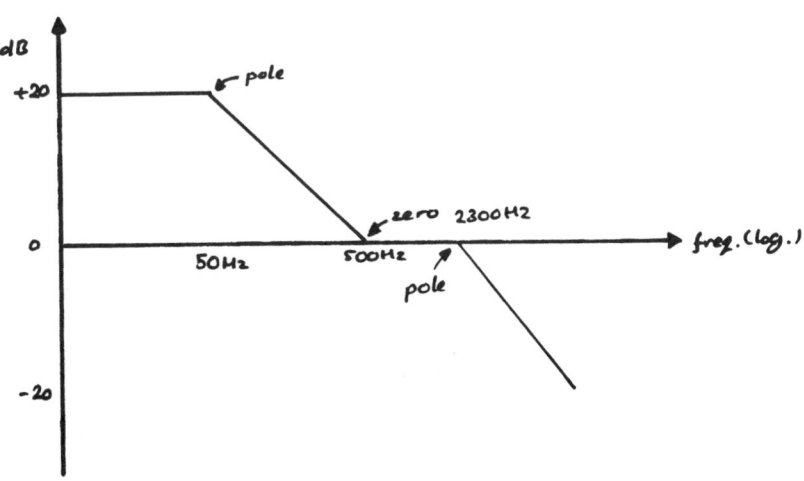

Figure 5.7 Approximate RIAA response

5.1.5 The Twin-T Notch Filter

This highly popular filter is one of the few RC filters to produce infinite attenuation at frequency $f_c = 1/2\pi RC$. It does this by producing phase shifts of $\pm \delta$ and adding these signals together. At f_c, δ becomes $\frac{1}{2}\pi$ (90°), and so the total phase shift, between the

209

signals in each T, is π (180°). This completely eliminates the signal at f_c. Such a filter gives a very sharp notch, and is often used to remove 50 Hz power line hum in audio systems. The use of a follower allows high resistances to be used, and so smaller high tolerance capacitors can also be used.

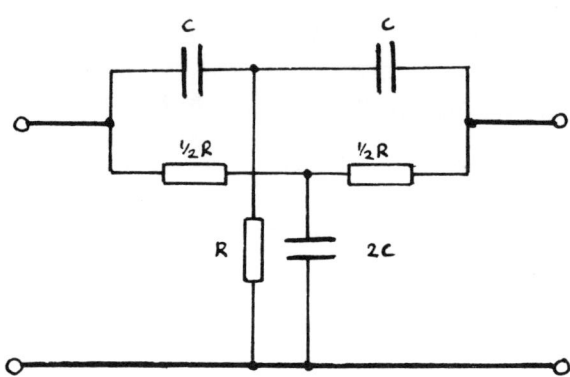

Figure 5.8 Twin T- Notch Filter

5.1.6 Active Filtering

I intend to give only a very brief discussion of active filtering, since this is a large topic best left to a dedicated book.

Filter Notation

I have already defined the terms: band pass, band stop (notch), high pass and low pass. A typical low pass filter characteristic is given on the next page.

The *pass band* is the part of the characteristic where significant attenuation does not occur. The edge(s) of the pass band is usually stated to be the frequency at which the response is 3 dB down, and out of the *ripple band*. This frequency is called the 'cut-off', 'critical' or 'corner' frequency, f_c, and the difference between the maximum and minimum passed frequencies is called the *bandwidth*, a term I have already been using. Many filters do not show a flat response in the pass band but, on a logarithmic plot, show a distinct ripple. The range of gain over which this occurs is referred to as the ripple band. An ideal filter has zero ripple band, but this is hard to produce. The *stop band* is defined to be the area where signal transfer becomes insignificant. It may be determined by quoting the frequency at which the attenuation exceeds some critical value, such as -40 dB (this represents a transfer of 1% of the input signal). The area between the pass band and stop band is called the *transition region* or *skirt*.

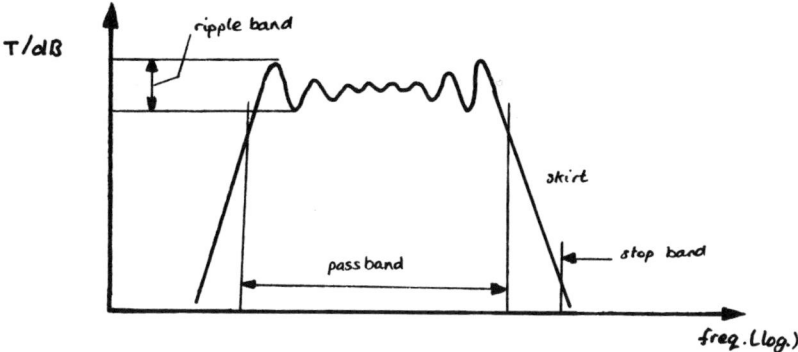

Figure 5.9 Filter nomenclature

It should be noted that the transfer function, **T**, of any circuit containing reactive components is, in general, complex, and both the gain, |**T**|, and phase shift, Arg **T**, are important quantities.

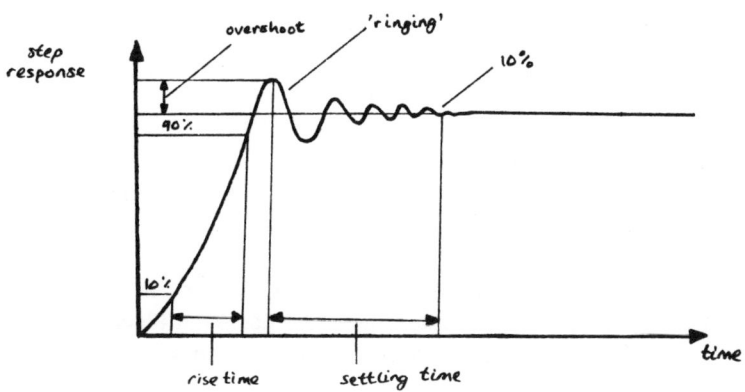

Figure 5.10 Step response nomenclature

Filters are often given names such as *Chebyshev* (pronounced 'sheby-sheff') or *Bessel*. These names refer to various polynomial functions, given the labels C_n and J_n for the examples, which the filter characteristic looks like. The subscript n refers the the order of the polynomials. Full details of these functions can be found in most books on Analysis or Differential Equations.

Another useful way of analysing filters is to study their response to a square wave input. Only the leading edge is studied.

The Sallen and Key Filter

This is essentially two cascaded RC filters, with an amplifier of gain k connecting the output to the 'ground' of the first RC stage. Often k is put equal to 1.

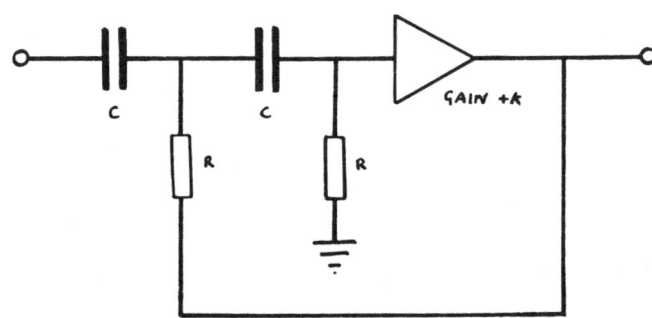

Figure 5.11 Sallen and key filter

The circuit is a 2-pole high pass filter; if you interchange the resistors and capacitors it changes to a 2-pole low pass filter.

5.2 Noise

This section is a short discussion of *noise*. Noise is a random unwanted signal superposed upon the required signal.

The amount of noise is often quoted in terms of the *signal to noise ratio*, which I shall define later. Since all electronic components generate noise, all electronic components deteriorate the signal. That is, they all increase the signal to noise ratio.

5.2.1 Spectral Distribution of Noise

The random nature of noise means that a precise theory cannot be produced to describe the noise signal generated by a component. We can, however, talk about the expected, or average, amplitude of this signal. It is not unreasonable to expect this amplitude to be dependent on frequency, and so we must talk in terms of the *spectral distribution* of the noise. When comparing the amounts of noise, we find it useful to talk in terms of the noise power, and the *power density spectrum* (p.d.s.) which gives the expected noise power per unit bandwidth. When combining noise we *always* add powers, *not* r.m.s. amplitudes

Mathematically, the noise power between frequencies f and f + df is given in terms of a noise power distribution function, which I am calling P(f).

$$dP = P(f)\ df \tag{5.1}$$

The Colour of Noise: White or Pink

The common forms of P(f) are referred to by the names *white noise* and *pink noise*. White noise is a noise with P(f) equal to a constant, pink noise has P(f) $\propto$ 1/f.

In other words, this means that for white noise the p.d.s. is such that there is an equal noise power per unit frequency bandwidth, regardless of where that bandwidth lies, and for pink noise it is such that there is an equal noise power per decade bandwidth, again regardless of where that decade lies.

5.2.2 Sources of Noise

The most significant sources of noise generated by components are resistance noise and shot noise.

Thermal Noise or Resistance Noise

The electrons in a conductor, in addition to a drift velocity due to the voltage across the material, have a random motion which is caused by thermal effects. This causes a random fluctuation in the p.d. measured across the conductor, i.e. a noise signal. Resistance noise is white.

The noise from a resistance, R, can be considered to be generated by a voltage generator of r.m.s amplitude V_n in series with an ideal noiseless resistor of magnitude R.

$$V_m = \sqrt{(4kTBR)} \tag{5.2}$$

B is the bandwidth over which the noise is measured, or of the circuit at whose output the noise appears, whichever is smaller. Here T, is the absolute temperature in Kelvin: $T/K = T/^{\circ}C + 273$.

If the component also has reactance we must replace R by the real part of Z.

Shot Noise

Shot noise is caused by charge carriers randomly crossing a p-n junction. This is a white current noise which is dependent on the current being conducted by the junction.

$$I_{sn} = \sqrt{(2qIB)} \qquad\qquad (5.3)$$

q is the electronic charge $= 1.6 \times 10^{-19}$ C.

The total current flowing across the junction is $I + I_{sn}$.

Flicker Noise

Flicker noise is due to fluctuations in the conductivity of semiconductor material. It is pink, and generally insignificant above 10 kHz.

Interference Noise

This is *stray*, externally generated, signals picked up by the circuit. Sources of interference noise include power supply ripple, induced voltages from changing magnetic fields/strong a.c. electric fields, poor switch contacts, etc.

Local radio stations can often cause radio frequency interference in sensitive equipment. Radio transmissions can be a particular problem. Anyone who has ever used an oscilloscope will also know how easy it is to get 50 Hz interference from the mains. This can make it quite hard to study radio frequency signals, unless a notch filter is fitted.

5.2.3 The Signal to Noise Ratio and Noise Figure

It should be emphasised that V_m and I_{sn} are, in fact, very small voltages and currents, of order μV and nA. The relative sizes of the noise signals and the desired signal can be expressed in terms of the signal to noise ratio. Simply stated, this is the ratio of signal power to noise power. Measuring the powers across the same load allows the SNR to be expressed in dB.

$$SNR/dB = 20 \log V_{signal}/V_{noise}$$

The noise figure of a circuit is the ratio of mean square noise at the output to the mean square noise at the input. This is also the ratio of the input SNR to output SNR

$$F = SNR_{input}/SNR_{output}$$

A circuit with a noise figure of less than 3 dB is a low noise circuit. Around 10 dB is the noise figure of an average circuit.

5.2.4 Reduction of Noise

The effective removal of noise really is dependent on the precise nature of the noise, and a spectrum analyser is a useful (if expensive) diagnostic tool. It is a good idea to find out whether the circuit is intolerably noisy before you spend money reducing noise. If your signals are a few volts, or more, you will be able to live with the noise in all but the most extreme cases.

A white noise interference on a narrow band signal can be mostly removed by the use of an amplifier with a very narrow bandwidth. R.F. interference on A.F. signals can be stopped if a low pass filter, with a critical frequency at about 100 kHz, is added. Mains interference can be cut with a very sharp notch filter. In such applications there is an advantage in using inductors without iron cores, since these are less likely to pick up local magnetic noise. Keeping the transformer well away from the circuit is often a good idea, and using toroidal transformers can cut down field leakage.

Other ways of removing R.F. interference are: shield the entire R.F. stage by putting it inside an earthed metal box, although this is not usually necessary; keep leads short and avoid loops of wire, which can act as a V.H.F. LC network; adding ferrite beads to long wires can cut out high frequencies.

Long twisted wires will often pick up the same noise signal, which can be simply removed using a differential input stage.

Crosstalk is the name for the phenomena where one part of a circuit, or transmission line, picks up interference from another part. This can be avoided by identifying the sources of crosstalk, and keeping them as far apart as possible. If necessary, they can be *shielded* by surrounding the offending circuit with a metal case which is earthed.

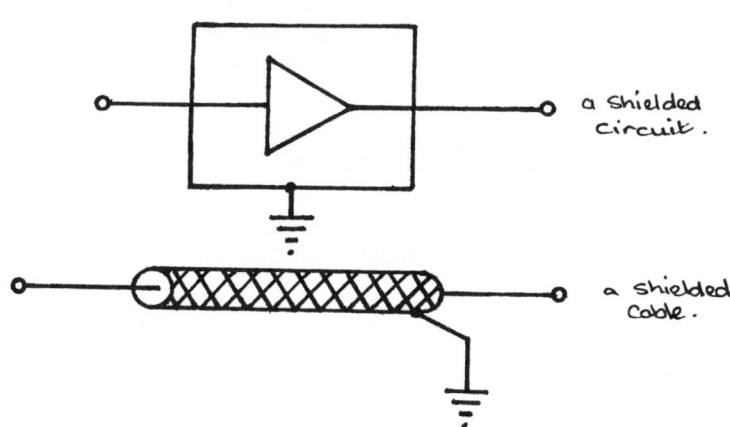

Figure 5.12 Shielding circuited and cables

215

Long wires are particularly prone to picking up noise, and this can be reduced by the use of shielded (co-axial) cables.

All this earthing can actually add problems of its own, due to earth currents being picked up as signals by other pieces of equipment. The *safe* technique is to connect all earth wires to the same terminal, as shown in the sketch.

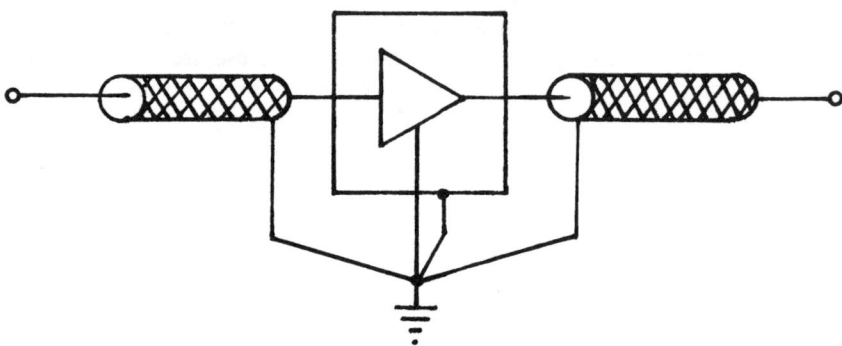

Figure 5.13 Avoiding earth line signals

The most noiseless amplifiers available are LASERs and MASERs. Many people are aware that LASER/MASER stands for Light/Microwave Amplification by Stimulated Emission of Radiation, but few are aware that these devices include the word amplification because they are actually used as amplifiers. The MASER is used as a microwave amplifier for radio telescopes, where the bandwidth is insignificant. Unfortunately such devices can only amplify at frequencies dependent on the energy levels of atoms/molecules, and these are set by nature not by circuit designers.

5.3 Power Output Stages

In chapter 2 I touched briefly on power output stages. I am completing the discussion, here since it is a subject which has widespread application in the field of audio frequency amplification.

5.3.1 Harmonic Distortion

The collector current of a bipolar power transistor is dependent on base-emitter voltage, and the dependence can be written $IC = f(V_{BE})$. To find $f(V + \delta V)$ we can use the Taylor expansion.

$$f(V + \delta V) = f(V) + \delta V\, f'(V) + \delta V^2\, f''(V)/2 + \ldots + \delta V^n\, f^{(n)}(V)/n! + \ldots$$

$f^{(n)}(V)$ signifies f differentiated n times at constant temperature, and evaluated at V. If the temperature is not constant this expansion does not hôld.

If δV is sufficiently small that terms in δV^2, or higher, can be neglected, we are left with the small signal model given in chapter 2.

$$f(V + v) \approx f(V) + v\, f'(V)$$

$f'(V)$ is identified with the small signal conductance g_m.

For a power transistor application v might be too large to neglect the term v^2, and so the expansion used must be:

$$I_C \approx I_{DC} + g_1 v_{BE} + g_2 v_{BE}^2 \qquad (5.4)$$

If v_{BE} is a sinusoid, $v_0 \sin \omega t$, then $v_{BE}^2 = v_0^2 \sin^2 \omega t$. Using the exponential forms of sin and cos, which I find decreases my dependence on data books, this can be easily evaluated:

$$
\begin{aligned}
\sin^2 \theta &= \{-\tfrac{1}{2}j(e^{j\theta} - e^{-j\theta})\}^2 \\
&= -\tfrac{1}{4}\{(e^{j\theta})^2 - 2e^{j\theta}e^{-j\theta} + (e^{-j\theta})^2\} \\
&= \tfrac{1}{2} - \tfrac{1}{4}(e^{j2\theta} + e^{-j2\theta}) \\
&= \tfrac{1}{2}(1 - \cos 2\theta)
\end{aligned}
$$

$\therefore \qquad I_C \approx I_{DC} + \tfrac{1}{2}g_2 v_0^2 + g_1 v_0 \sin \omega t - \tfrac{1}{2}g_2 v_0^2 \cos 2\omega t \qquad (5.5)$

This means that, to a signal which is a pure sinusoid at frequency $2\pi\omega$ has been added an unwanted signal at frequency $4\pi\omega$ (i.e. exactly twice the frequency). Such a signal is said to be the *second harmonic* of the signal, and the effect is known as *harmonic distortion*. As shown in figure 5.14, the effect can be quite dramatic.

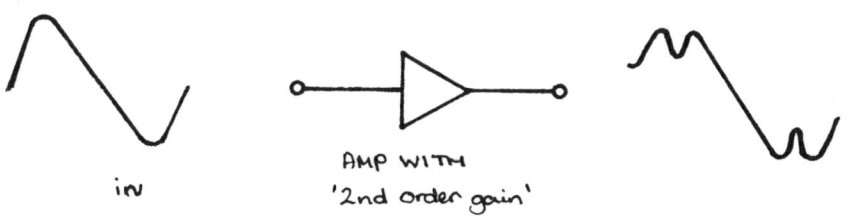

in
AMP WITH
'2nd order gain'

Figure 5.14 The effect of second order gain

If the signal was so large that terms in v^3, v^4, ... had to be included, there would also be distortion due to the introduction of the third (3ω), fourth (4ω), etc. harmonics.

The amount of nth harmonic distortion is usually determined from the ratio:

amplitude at frequency nf/that at frequency f

The total harmonic distortion is also simply determined.

$$\text{T.H.D.} = \sqrt{(\Sigma \ A_n^2)}/A \qquad (5.6)$$

A_n is the amplitude of the nth harmonic of the input frequency, and A is the amplitude of the input frequency.

A similar technique can be applied to any d.c. amplifier which is made to amplify a signal which consists of a large biasing (D.C.) signal and a small a.c. signal. The response of the amplifier is, therefore, written:

$$I = I_{DC} + g_1 v_{in} + g_2 v_{in}^2 + \dots + g_n v_{in}^n + \dots \qquad (5.7)$$

The series must converge, which means that as $n \rightarrow \infty$ the amount of nth harmonic distortion must approach zero. Of course, the amplifier need not be a transconducting amplifier; the method is equally valid for a voltage amplifier or a current amplifier.

The values of the constants, g_n, may be calculated from an accurate equation (if known), derived from a plot of the amplifier transfer functions, or measured using a spectrum analyser or similar device.

Another quantity used to compare power amplifiers is the *collector efficiency*, η.

$$\eta = \text{max a.c. o/p power/d.c. power from PSU} \qquad (5.8)$$

5.3.2 Class A Output Stages

A class A power amplifier is an amplifier which gives large power gain, and is always biased into conduction.

Emitter Follower Type

The simple follower above is an example. The use of a PNP follower and NPN follower cascade mostly removes the voltage drop by the output transistor base-emitter junction. The current source must be big enough to allow the output voltage, in minimum load conditions, to reach an acceptable maximum. If it is not, then clipping will result.

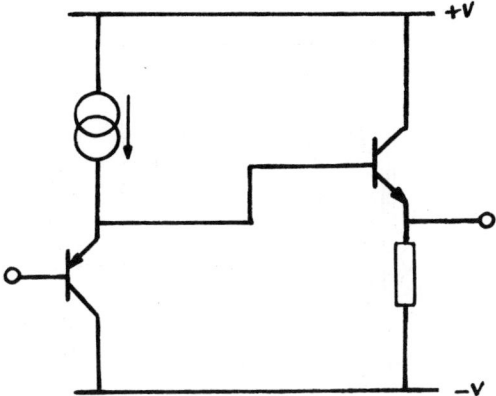

Figure 5.15 Emitter follower power amplifier

Either a split supply can be used, if the input is centred on 0 V, or the circuit must be biased such that the quiescent emitter voltage, of the output transistor, is approximately half-way between the power supply voltages.

Class A Push-Pull

An alternative circuit can be used to give a lower harmonic distortion. The operation is the input stage has two outputs which are output of phase by π (180°). These are applied to two identical power amplifiers, and the outputs applied to either end of the output transformer.

If the transistors are identical, their behaviour is described by the same power series.

$$I_C = I_{DC} + g_1 v_{BE} + g_2 v_{BE}^2 + \ldots$$

Since the amplifiers are driven out of phase by π then $v_{BE2} = - v_{BE1}$.

i.e. $\qquad\qquad I_{C1} = I_{DC} + g_1 v_{BE} + g_2 v_{BE}^2 + \ldots$

and $\qquad\qquad I_{C2} = I_{DC} - g_1 v_{BE} + g_2 v_{BE}^2 - \ldots$

therefore, the primary current is:

$$I_P = 2g_1 v_{BE} + 2g_3 v_{BE}^3 + \ldots \qquad\qquad (5.9)$$

All the even harmonics have been eliminated, and the T.H.D. of the amplifier is much lower.

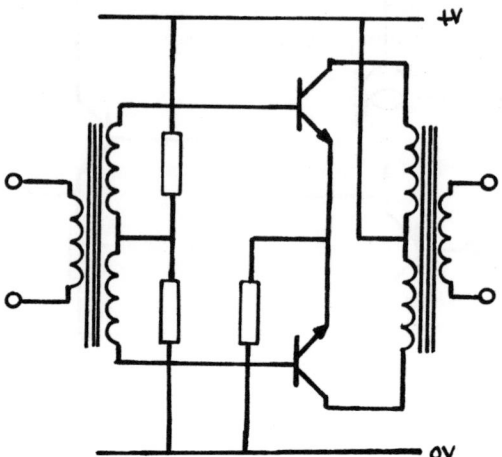

Figure 5.16 Class A push-pull amplifier

General Remarks

Because the transistors are always on, class A amplifiers are power wasters. The theoretical maximum collector efficiency is only 50%. The last two circuits suffer from the problem that their output impedance is purely reactive, and so is proportional to the frequency.

5.3.3 Class B (and AB) Output Stages

A class B output is one in which there is no bias applied to the output stage transistors. A class AB is one in which the output transistors are just biased into conduction.
The complementary emitter-follower is a widely used class B stage. (See Figure 2.29)

As previously discussed, it does suffer from crossover distortion. A simple solution is the diode network discussed earlier. The network is not ideal because it is not thermally stable. As transistors get hot, at constant current, the base-emitter drop decreases by 2.1 mV/K. If the diode temperature is constant this change appears as a small signal across the base-emitter junction and the transistors are biased further into conduction. This increases the collector current and the transistors get hotter. Positive feedback occurs and, if you are not careful, the transistors may be destroyed. By thermally coupling the diodes to the power transistors this effect can be reduced.

The thermal stability can be further enhanced by putting two small resistors, usually a few Ω, between the transistor emitters and the output. If the p.d. between the bases is fixed, and the temperature of the devices increases, the base-emitter drop will decrease and the transistors will be biased slightly further into conduction. The increase in quiescent emitter current causes a larger drop in the emitter resistors, and so lifts the emitters and decreases the bias.

Extra forward bias voltage is required because of the emitter resistors. This is often provided by a small resistor or an extra diode.

General Remarks

Class B complementary emitter-followers are also push-pull amplifiers, but unlike the class A push-pull, it does not eliminate the even harmonic distortion. With bad biasing the crossover distortion can also be a problem.

The points in favour of the circuit are its 80% maximum collector efficiency, it's simplicity, and it's purely resistive output impedance. High power output stages, such as in hi-fi power amplifiers, will be class B (or AB) almost without exception.

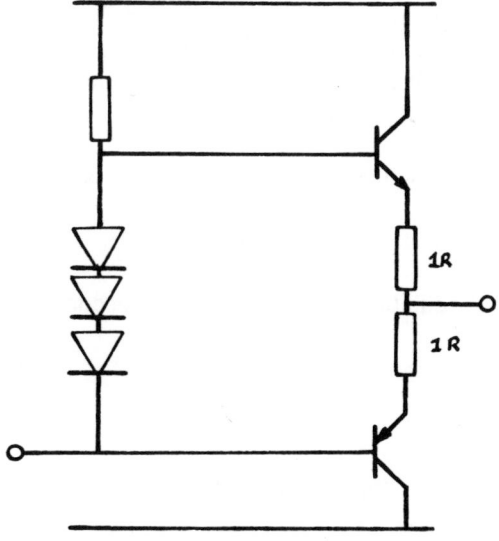

Figure 5.17 Temperature compensated class B amplifier

5.4 Sinusoidal Oscillators

This section is a discussion of sinusoidal oscillators. The discussion of square wave (relaxation) oscillators is in chapter 4.

5.4.1 Amplifiers and Oscillation

In chapter 3 I derived a formula for the gain of any negative feedback amplifier. In general, most amplifiers, and feedback networks, have some reactive element, and so a more realistic formula is:

$$A' = \frac{A_o}{1 - A_o\beta} \tag{5.10}$$

The reason why this equation includes $-A_0\beta$, unlike that derived in chapter 3, which includes $+A_0\beta$, is because in chapter 3 the feedback was assumed to be negative whereas here the *sign* of the feedback is incorporated into the variable β .

Since $A_0\beta$ is frequency dependent, there may be some frequencies at which $A_0\beta$ = 1. Substituting in this condition gives the result that the closed loop gain is infinite. Using $A' = V_{out}/V_{in}$ we deduce that an "infinite gain" means that there is a finite output for zero input at any frequency for which $1 - A_0\beta$ = 0. This means that the amplifier acts as a source of signals at those frequencies, which is a free running oscillator. For a single sinusoidal output, there must be only one frequency at which the condition is satisfied.

The condition to produce oscillations may be written in polar form. (n is any integer)

$$|A_0\beta| e^{j \text{ arg } A_o\beta} = e^{2jn\pi} \tag{5.11}$$

Therefore, to produce oscillations at a frequency f_0, the conditions that must be satisfied are:

$$|A_0(f_0)\beta (f_0)| = 1$$

and $\qquad\qquad\qquad\qquad\qquad\qquad\qquad\qquad\qquad\qquad$ (5.12)

$$\text{arg } A_0(f_0)\beta (f_0) = 0, 2\pi, 4\pi...$$

That is, the magnitude of the loop gain must be one, and the total phase shift of the amplifier and feedback network must be zero (or 360° etc).

If these conditions are satisfied in an amplifier then it will oscillate. Such an effect in a signal amplifier is described as *instability*, and must be avoided in a useful amplifier. If oscillation is a serious problem then one, or both, of the conditions (5.12) must be forced untrue at the frequency of oscillation. This is the reason why frequency compensation is used in operational amplifiers. The internal, or external, frequency compensation network acts to drop the gain at high frequencies, which is where such oscillations can occur, to such a low level that the amplitude of oscillation is insignificant, or even zero.

5.4.2 The Wien Bridge Oscillator

For the inverting input: $\qquad V_-/V_o = R_i/(R_i + R_f)$

and at the non-inverting input:

$$V_+/V_o = \{R||(1/j\omega C)\}/\{R + 1/j\omega C + R||(1/j\omega C)\}$$

This can be simplified: $\qquad V_+/V_o = R/\{3R + j(\omega R^2 C - 1/\omega C)\}$

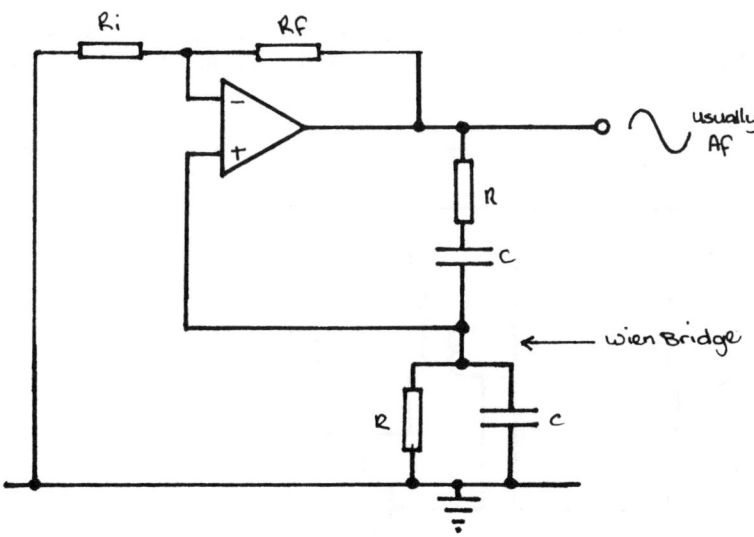

Figure 5.18 Wien bridge oscillator

For an ideal amplifier we must have $V_+ = V_-$, and so V_+/V_o must be a wholly real quantity.

$\therefore \qquad\qquad\qquad\qquad \omega R^2 C = 1/\omega C$

$\Rightarrow \qquad\qquad\qquad\qquad \omega = 1/2\pi RC \qquad\qquad\qquad\qquad (5.13)$

also
$$R_i/(R_i + R_f) = \tfrac{1}{3}$$
(5.14)

Equation (5.14) can be satisfied by putting $R_f = 2R_i$.

5.4.3 Tuned Oscillators

A small signal amplifier which has a very narrow (adjustable) bandwidth is called a *tuned amplifier*. A simple circuit can be made with a parallel LC circuit.

At the resonant frequency, f_0, of the LC network the gain, $-g_m Z_L$, reaches a maximum. Except for a narrow band around f_0 the gain is insignificant.

This circuit can be simply turned into an oscillator by replacing the inductor with a transformer. The primary coil acts as the inductor in the LC network, and the secondary is used to feed the output back to the input. Since the transistor provides a phase shift of π (inverting configuration) oscillations will only occur if there is a phase shift of π introduced by the transformer.

An analysis of this oscillator (valid at audio frequencies) can be made by substituting in the small signal model. The internal resistance of L_b is much less significant than that of L_c.

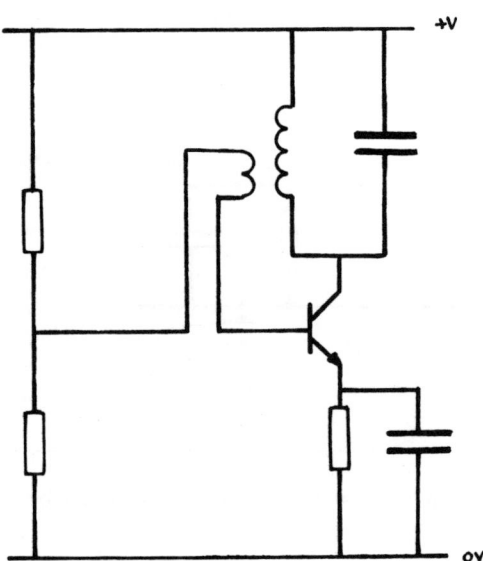

Figure 5.19 Tuned collector oscillator

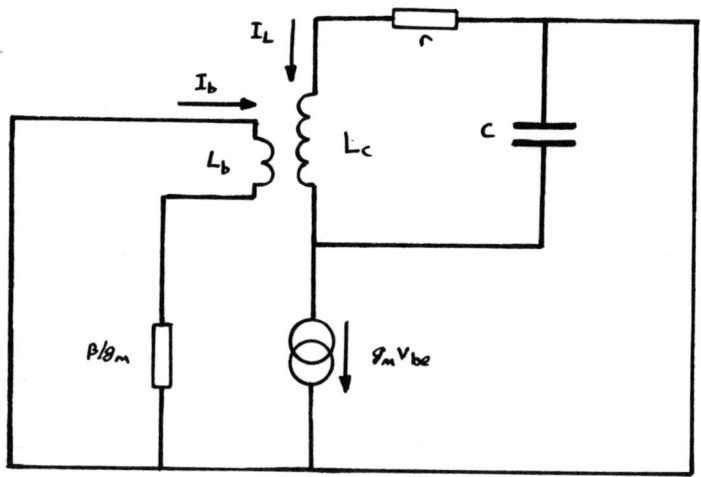

Figure 5.20 Small signal equivalent circuit of tuned collector oscillator

Applying Kirchoff's second law to the collector loop:

$$i_L(r + j\omega L_c) + j\omega M i_b - (g_m v_{BE} - i_L)/j\omega C = 0$$

This equation contains i_b, and so, for convenience, I shall use $\beta i_b = g_m v_{BE}$.

$$\Rightarrow \qquad i_L r + j\omega L_c i_L + j\omega M i_b - \beta i_b/j\omega C + i_L/j\omega C = 0$$

collecting terms:

$$i_L(r + j\omega L_c - j/\omega C) + i_b(j\omega M - \beta j/\omega C) = 0$$

or $\qquad i_L(j/\omega C - r - j\omega L_c) = i_b(j\omega M - j\beta/\omega C)$ $\qquad$ (5.15)

And for the base loop:

$$i_b(\beta/g_m + j\omega L_b) + j\omega M i_L = 0 \qquad (5.16)$$

Combining these equations gives equation (5.17).

$$-r\beta/g_m - j\omega r L_b + j\beta(1/\omega C - \omega L_c)/g_m - L_b/C + \omega^2 L_b L_c = \omega^2 M - \beta M/C \quad (5.18)$$

The RHS of this equation is wholly real, and therefore:

225

$$-\omega r L_b + \beta/\omega C g_m - \omega L_c \beta/g_m = 0$$

$$\Rightarrow \qquad \omega^2 = 1/(L_c C + L_b Crg_m/\beta) \qquad (5.19)$$

If $r \ll \beta/g_m$ then this result may be simplified:

$$f \approx 1/2\pi\sqrt{(L_c C)}$$

A similar circuit can also be built around a FET.

5.4.4 Precision Oscillators

Simple oscillators do not oscillate at a precisely determined frequency (due to component tolerance), and also the impedances/amplifier parameters tend to change with temperature. For accuracy, such circuits often have to be left running for a few hours before the frequency stops drifting. Piezo-electric crystals, which were mentioned earlier, are used to fix the frequency in such situations. The temperature coefficient of a crystal oscillator is much less than that for conventional oscillators, and in very precise oscillators the crystal is kept at constant temperature in an oven. The crystal has two resonant frequencies. At the lower, it behaves as a series LCR network, and at the higher, as a parallel LCR network.

Questions for Chapter 5

5.1   Show that, for the RC network of figure 5.i the 'gain' of this 'amplifier' can be written:

$$|A| = 1/\sqrt{(1 + \omega^2/\omega_C^2)}$$

ω_C is the corner frequency (3dB point) of the network.

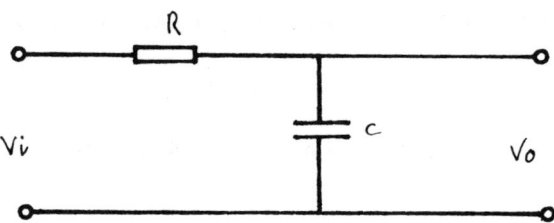

Figure 5.i An 'amplifier'

5.2   An operational amplifier is compensated so that it has a 3dB point at 1 kHz. Use the answer of question 5.1 to find an expression for the open loop gain of the amplifier at all frequencies. Show that the application of negative feedback increases the bandwidth from 1 kHz to some higher value. By what factor is it increased?

5.3   Figure 5.ii shows a *Wien bridge* network. Find an expression for the transfer function of the network. What is its:

i   Maximum 'gain'; and
ii   Phase shift at maximum gain.

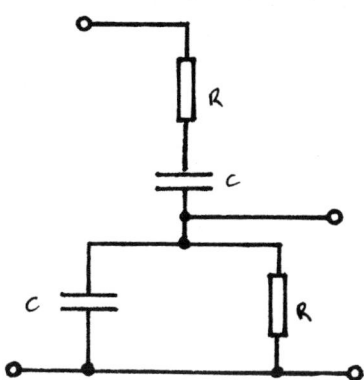

Figure 5.ii The Wein bridge

5.4   Compare and contrast the effectiveness of an LC, Wien bridge, and Twin-T notch filter in cutting mains interference at 50 Hz.

5.5   An attenuator is made from a resistor, R, and a NIC'd capacitor, C. Find an expression for the 'gain' and phase shift of this network.

5.6   A 47 kΩ resistor, which is part of a very low current amplifier, is found to be unacceptably noisy. If 5 nA flows through the resistor find the signal to noise ratio as measured across the resistor at 290 K. It is suggested that this problem can be reduced by cooling the circuit with liquid nitrogen. Find the signal to noise ratio at 70 K. The bandwidth of the 'scope used to measure the noise is 2 MHz.

5.7   Draw a noise equivalent circuit for a bipolar transistor.

5.8   Show that the collector efficiency of a class A power amplifier is 50%.

5.9   A class A power amplifier is described by the 2nd order expression below:

$$I = A + BV + CV^2$$

What is the percentage 2nd. harmonic distortion?

5.10   (*harder*) By Fourier analysis of a half-wave rectified signal calculate the maximum collector efficiency of a class B power amplifier. You may assume that terms in the 2nd. harmonic or higher are insignificant.

5.11   A Wien bridge oscillator is made with the Wien network constructed, as below, from two resistors, R and S, and two capacitors, C and D. Find an expression for the oscillating frequency of the oscillator.

5.12   Find an expression for the oscillating frequency of a tuned drain oscillator.

Answers to Numerical Problems

5.3 $\frac{1}{3}$; 0 radians

5.6   16 dB; 22 dB

5.10   80%

6

RADIO AND DATA TRANSMISSION

6.1 The Transmission of Data

6.1.1 Signals, Carriers and Modulation

Signals and Carriers

A *signal* is any information which has to be transmitted. For example, if we are interested in the voltage measured across a microphone, then this a.c. waveform is the signal. The *carrier* is the medium in which the signal is transmitted. In the above example, the carrier is the cable through which the current flows. Basically, any medium in which information can propagate through space may be used as a carrier. Common examples are: air; metallic conductors; light; radio waves; etc. However, more precisely, we should recognise that in any situation where waves may exist we can use the waves as a carrier for the information.

Mathematically, a travelling wave is a disturbance which propagates through a coordinate space. Any wave which moves to the left may be written:

$$g(x - ct + \varphi)$$

where x is the instantaneous position of the wave, c is the wave speed, φ is a phase factor and t is time. For example, consider the simple sinusoidal wave sin (x - t).

Any travelling wave, such as that above, or standing wave, is a solution of a special *partial differential equation* called the *wave equation*.

$$c^2 g_{xx} = g_{tt}$$

A study of waves and the wave equation is not really possible at this level.

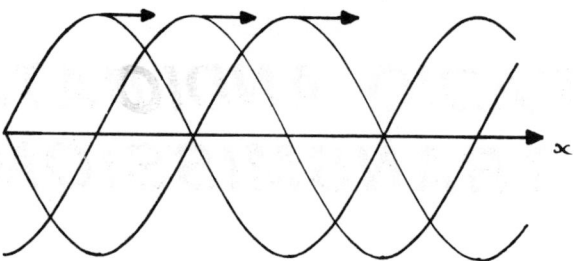

Figure 6.1 A sinusoidal wave

Modulation

Modulation is the general name for the process by which we make the signal travel along/through/with the carrier. Consider the simple sinusoidal wave, which is easy to create and easy to study. There are three properties of a sinusoid which completely specify its form.

amplitude;

frequency; *and*

phase.

If any of these quantities are modified in a regular manner, and to an extent determined by the instantaneous value of some signal, any receiver which can detect these quantities may decode (demodulate) the carrier and recover the signal.

6.1.2 Some Modulation Methods

The phase is not often used in the modulation of sinusoidal waves because of the difficulty in determining the actual phase of the wave at a remote receiver.

Amplitude Modulation

A system of modulation in very widespread use is amplitude modulation. Here the amplitude of the transmitted (carrier) wave is made *linearly dependent* on the instantaneous value of the signal, which is often a voltage.

i.e.
$$(1 + bV_{sig}) \sin (kx - \omega t)$$

ω = angular frequency of wave, $k = 2\pi/\lambda$ which is called the wave number (λ = wavelength). The wavespeed is a property of the carrier, and usually a constant of value ω/k.

Usually V_{sig} is itself a sinusoid of much lower frequency, and the wave has the familiar form of figure 6.2.

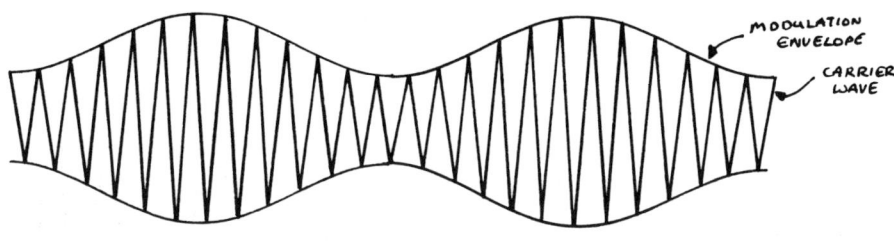

Figure 6.2 Amplitude modulation

With this system it is possible for the receiver to decode the signal to give the absolute frequency of the modulation and the relative amplitude of the modulation. This data is sufficient for good, but not high-fidelity, transmission and reception of sound. The A.M. system is prone to *noise*.

Frequency Modulation

Frequency modulation is a very high quality data transmission system. To appreciate this you only have to listen to an A.M. radio station (medium wave) followed by an F.M. radio station (V.H.F.). In this system the actual frequency of the carrier wave is made linearly dependent on the instantaneous value of the signal.

Although simple to conceive, this system is technically much harder to use, which is why A.M. stations have been in operation almost since the invention of the first radio set, but F.M. stations are only a recent invention.

The F.M. system is immune to amplitude noise, because the demodulator ignores the amplitude completely, and so can be used for high-fidelity transmission of sound.

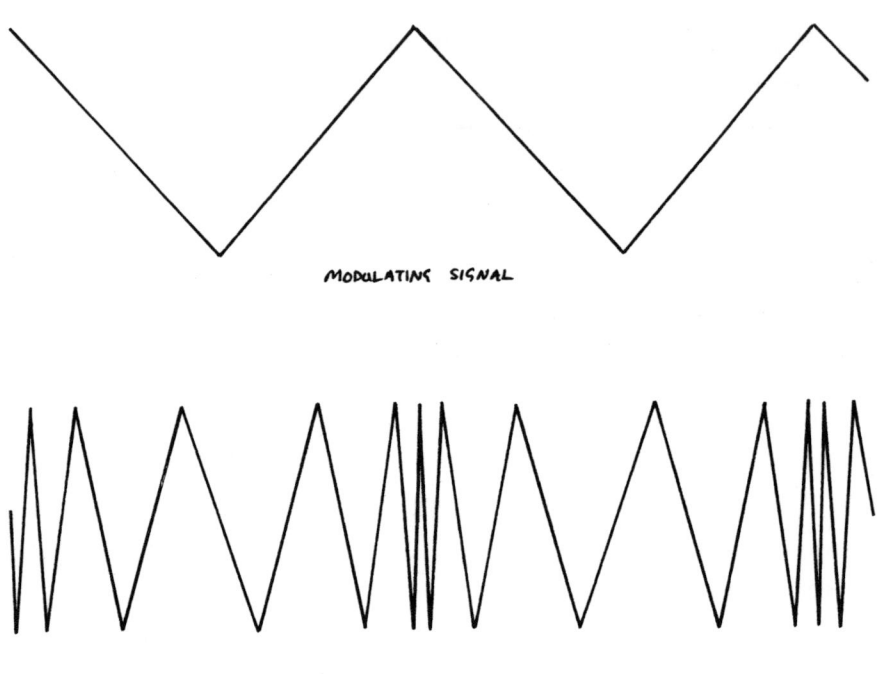

MODULATING SIGNAL

F.M. MODULATED SIGNAL

Figure 6.3 Frequency modulation

Carrier Keying or Morse Code (C.W.)

The earliest radio/telegraphic transmission was done with a basic form of modulation known as carrier keying. The most common form uses the *Morse code* system.

In CW the actual carrier is turned on and off in bursts. This is easy to transmit and easy to detect. However, it cannot be used to transmit hi-fidelity sound because it is a digital system with a very low sample frequency. The digital transmission system does mean that it is highly immune to noise.

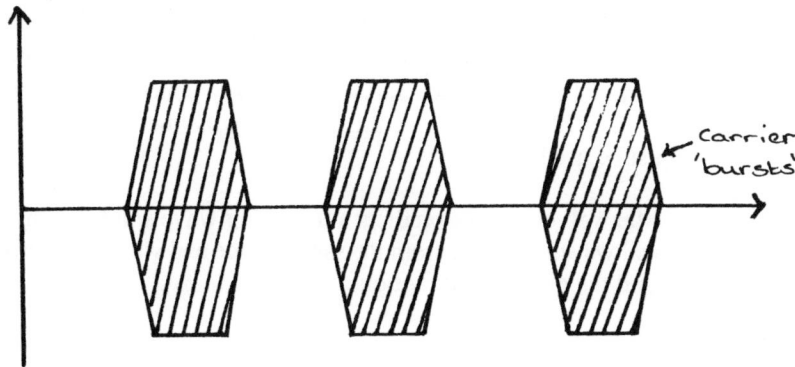

Figure 6.4 Carrier keying

Pulse Width Modulation (P.W.M.)

P.W.M. is, in essence, modern C.W. The length of the carrier pulses are made dependent on the signal, which means that the signal has to be sampled at regular intervals.

If the sample rate is something like 50 kHz there is little loss of quality, due to sampling, at audio frequencies. There is also an improvement in signal quality because of the inherent noise immunity of the digital transmission method.

6.2 Radio

6.2.1 Maxwellian Electromagnetism and Hertz's Experiment

Radio Waves

Maxwell's equations are a set of four equations which completely describe classical electromagnetism. When Maxwell introduced the equations he also showed how they could be manipulated to give wave equations for the electric and magnetic fields. Such *radio* waves are generally given one of the following names, according to the frequency:

longwaves	<	1 MHz	
shortwaves	<	10 MHz	
microwaves/radar waves	<	10 GHz	$(10^{10}$ Hz)
infra-red light	<	10^{14} Hz	
ultra-violet light	>	10^{15} Hz	
X-rays	>	10^{17} Hz	
γ-rays	>	10^{20} Hz	

Visible light is a very narrow band between approximately 4×10^{14} Hz and 8×10^{14} Hz. X-rays are sometimes called *deceleration radiation* since they are emitted when high energy electrons are stopped in a target such as iron. γ-rays are emitted as part of the decay of unstable atoms. In electronics we are only concerned with the waves which have frequencies < 100 GHz. Above this frequency the methods of optics, rather than electronics, must be used.

An analysis of the Maxwell equations predicts a travelling wave solution which has the magnetic and electrical waves radiating from the same source and which are perpendicular to both themselves and the direction of motion.

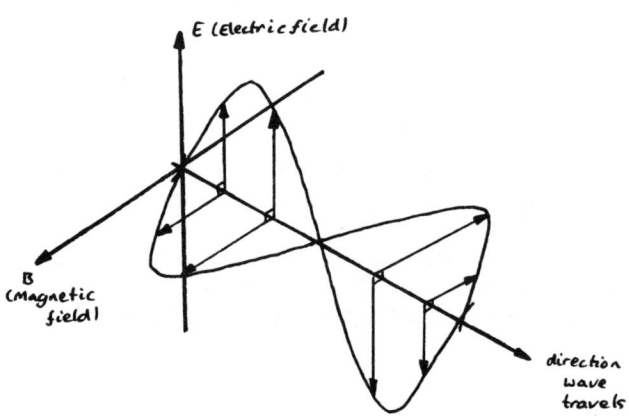

Figure 6.5 Electromagnetic waves

You cannot produce the electric wave without the magnetic wave, and vice-versa. It is easily shown that the electrical wave usually has a larger amplitude than the magnetic wave. The *polarisation* of the wave is the direction of the electrical field relative to some arbitrary reference. This is usually the ground.

Both electrical and magnetic fields can interact with electrons, and so induce a current in an appropriate transducer. Most transducers detect either the electrical or the magnetic fields, but not both. For maximum detected intensity the *active axis* of the transducer must be either parallel or anti-parallel with the direction of the appropriate field line. The active axes of transducers to detect the electric and magnetic fields are shown in Figure 6.6.

Most radio stations transmit with vertical polarisation, and so signals can be received with either a vertical wire or horizontal coil.

234

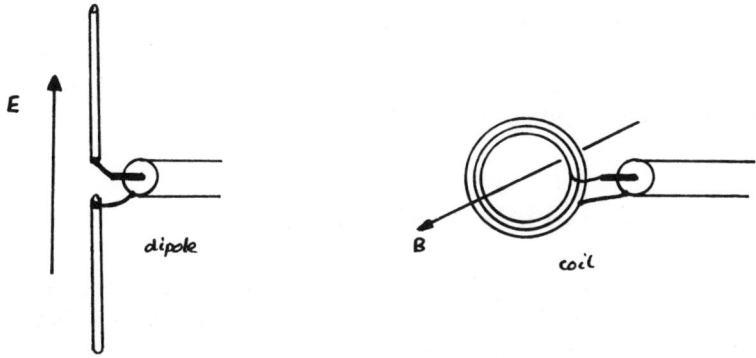

Figure 6.6 Dipole and coil antennae

Hertz's Experiment

Hertz, after whom the unit of frequency is named, did the first experiments to check Maxwell's prediction. He used an LC network which when *struck*, by applying a voltage, would *ring* like a tuning fork, but emitting radio frequency waves instead of sound. This was detected in a single coil. When the induced current in the coil was large enough a spark was created in the gap.

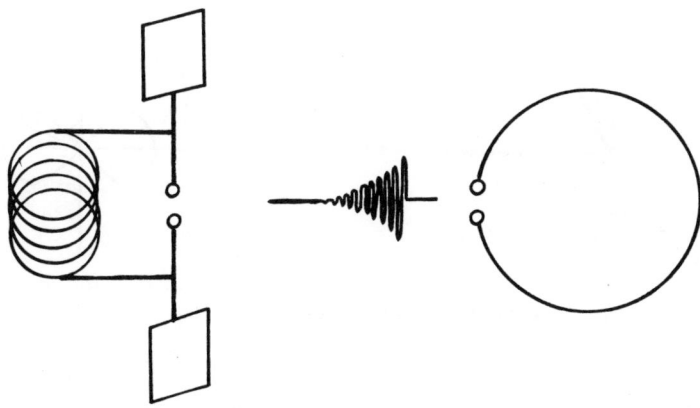

Figure 6.7 Hertz's experiment

When Hertz struck the LC network a click could be heard, due to a spark in the receiver gap. Maxwell's theory predicted that the waves would pass through paper, but be reflected from metal, and simple experiments could be done to verify these predictions. At first, Hertz could only transmit over small distances but, after Marconi demonstrated the transmission and reception of radio waves across the Atlantic, it was not long before wireless telegraphy was in widespread use.

6.2.2 More on Aerials and Antennae

The usual form of the transmitting aerial is the dipole. This is two wires, or rods, which are driven by an R.F. a.c. signal connected as shown in figure 6.8. For maximum *effective radiated power* output, at a particular frequency, the aerial is constructed such that its natural frequency is at that frequency. This criterion demands that the total length of the dipole must equal half of the free-space wavelength of the radiation. The wavelength is calculated using $c = f \lambda$.

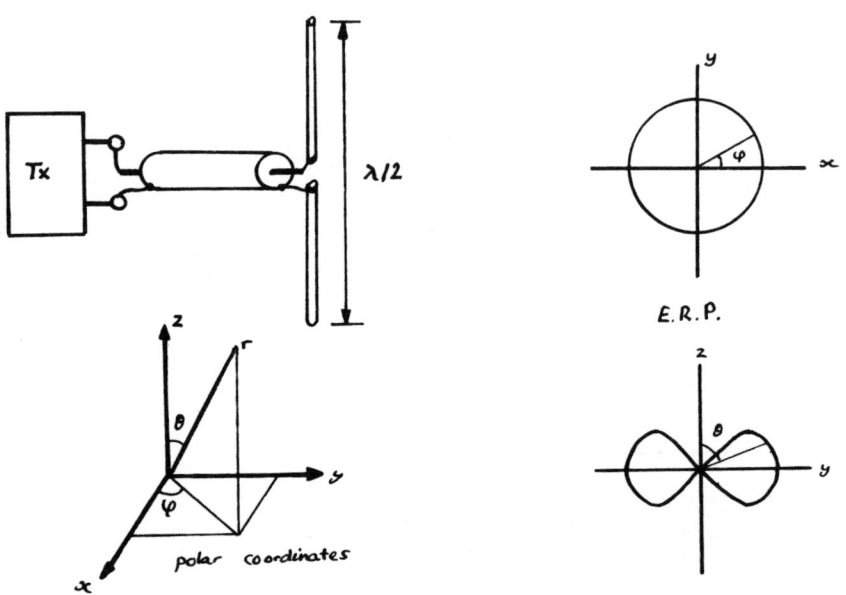

Figure 6.8 How ERP depends on direction and tuning a dipole

A single dipole radiates the same E.R.P in all horizontal directions, but the E.R.P. tends to zero as the angle between the transmitting dipoles axis and the active axis of the receiver approaches $90°$ and $270°$. The E.R.P. can be plotted as a surface in 3d, but since the dependence on the polar angle will usually be $\sin^2 \theta$ this is ignored and only the dependence on the equatorial angle is shown.

For broadcast transmission this is the best solution, since the signal must be received by a set at any angle to the transmitter. However, we often want more narrow sources so that the signal is only sent in a particular direction. To produce beamcast transmissions *reflectors* and *directors* are used. A reflector is a rod placed $\frac{1}{4}\lambda$ behind the dipole and directors are rods placed at integral multiples of $\frac{1}{4}\lambda$ in front of the dipole. These rods act as *passive radiators*, which means that currents are induced in them due to the fields from the dipole and those currents cause the rods to radiate, and the interference between the different sources acts to intensify the radiated field in the direction that the dipole is pointing, and to attenuate it in all other directions. This sort of antenna is also used as a receiver aerial for F.M. or T.V. transmissions. Here, the directional dependence is used to pick out signals from a particular station whilst ignoring others transmitted at the same frequency, but in different directions. Such receiver aerials offer better performance than simple loops or wires but have the disadvantage that their dimensions are of the order of a few wavelengths. This is acceptable at V.H.F. and U.H.F., where the wavelengths are typically 1 m, but are totally useless for long or medium wave transmissions where wavelengths can reach a few kilometres. For long and medium wave transmissions the aerial is usually a coil wound around an iron rod.

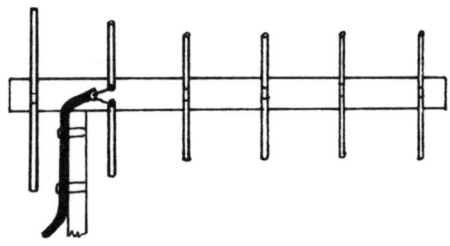

Figure 6.9 A highly directional aerial

6.2.3 The A.M. Signal

Before discussing circuits to receive and demodulate A.M. broadcast transmissions I need to make an important point about A.M. signals.

At first glance one would expect that the A.M. signal:

$$(1 + V_o \sin \omega_m t) \sin \omega_c t,$$

in which modulation at frequency ω_m is transmitted using a sinusoidal R.F. carrier at frequency ω_c, contains only the carrier frequency ω_c. The A.M. signal, in fact, contains three sine waves. The spectrum of the signal above is shown in Figure 6.10.

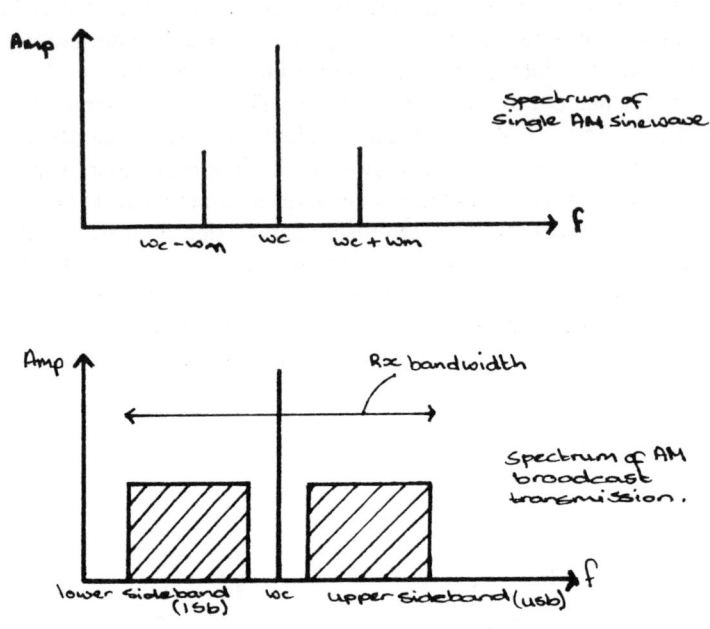

Figure 6.10 Spectrum of the A.M. signal

If the receiver input stage does not pass all three frequencies the received signal will not have the correct A.M. *shape* and there will be considerable distortion. For reception and successful demodulation of sound, the pass band of the input stage must be twice the audio bandwidth, which is approximately 20 kHz.

6.2.4 The Crystal Set

To receive a radio transmission we must make a transducer which can detect radio waves, and tune that transducer so that we only pick up transmissions at one particular carrier frequency. This can be accomplished using an LC network which is *tuned* such that its resonant frequency is equal to that of the transmission.

The A.M. signal is demodulated by putting it through a half-wave rectifier, and smoothing the output. This leaves a d.c. voltage with the audio signal superposed upon it. A high-pass filter cuts out the d.c., and the a.c. is dropped across an earpiece.

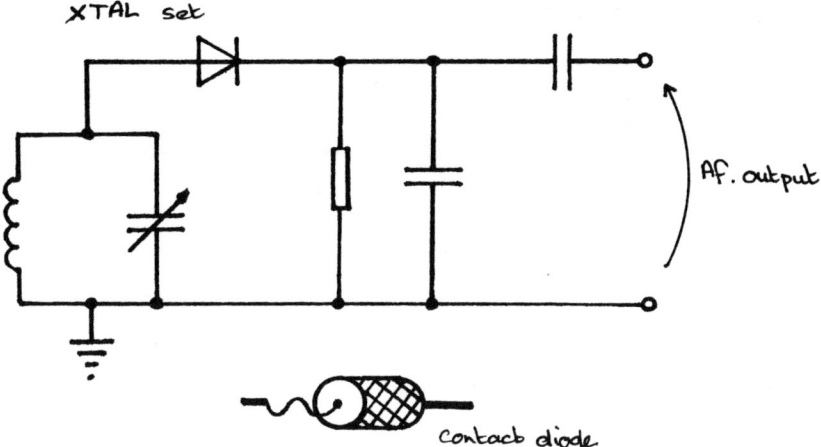

XTAL set

Af. output

Contact diode

Figure 6.11 Crystal set

In this circuit a germanium signal diode is used to minimise the forward voltage drop of the rectifier. Early crystal sets used a diode formed by fusing a small wire (called a whisker) with a semiconducting crystal, to form a rectifying junction. This is the derivation of the circuit's name. This type of diode is called a *point contact* diode.

6.2.5 The Tuned Radio Frequency Receiver (T.R.F. Set)

The big problem with crystal sets is getting a large enough signal to forward bias the diode. The T.R.F. set uses a tuned amplifier to give gain around the carrier frequency, and this allows the reception of much weaker signals. The same demodulator circuit is used, but this is usually followed with audio amplifiers.

For even greater sensitivity, a number of tuned stages may be used. When picking components for this sort of circuit care must be taken that the transistors used behave properly at radio frequencies. The data to look for is the transition frequency, f_T, which gives the frequency at which the current gain goes to 0 dB. Below this frequency the current gain obeys the gain-bandwidth relationship, but at low frequencies it approaches the d.c. gain. The bandwidth, B, is defined to be the frequency at which the current gain is 3 dB down from the d.c. gain.

if f « B $\beta f = f_T$

(6.1)

if f » B $\beta = \beta_{DC}$

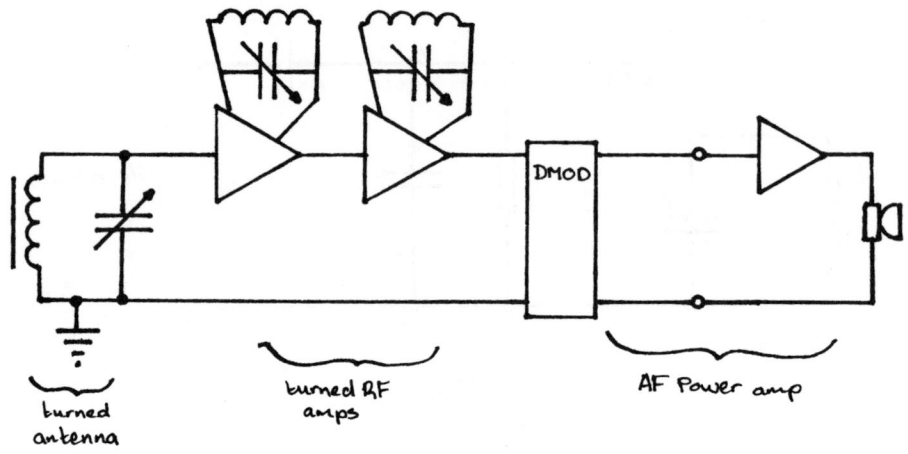

Figure 6.12 TRF set

6.2.6 Non-Linear Gain and Intermodulation

The small signal gain of any active device is represented by a power series, in terms of the input. Normally, the terms in v_{in}^2 or higher are neglected, since they are insignificantly small. In radio circuits this non-linear gain is exploited.

Consider an amplifier with second order gain, to which two signals are applied.

i.e. $$i_o \approx g\, v_{in} + h\, v_{in}^2$$

and $$v_{in} = v \sin \omega t + w \sin \Omega t$$

$\Rightarrow$ $$i_o = gv \sin \omega t + gw \sin \Omega t$$

$$+ h(v \sin \omega t + w \sin \Omega t)^2$$

$$= gv \sin \omega t + gw \sin \Omega t$$

$$+ hv^2 \sin^2 \omega t + hw^2 \sin^2 \Omega t$$

$$+ 2_h vw \sin \omega t \sin \Omega t$$

240

$$\Rightarrow \quad i_o = gv \sin \omega t + gw \sin \Omega t$$

$$+ \tfrac{1}{2}hv^2(1 - \cos 2\omega t) + \tfrac{1}{2}hw^2(1 - \cos 2\Omega t)$$

$$+ hvw \sin (\omega + \Omega)t + hvw \sin (\omega - \Omega)t$$

The spectrum of this signal has undergone quite a dramatic change as a result of amplification. There have been sum and difference frequencies introduced, as well as the 2nd. harmonics of the input frequencies.

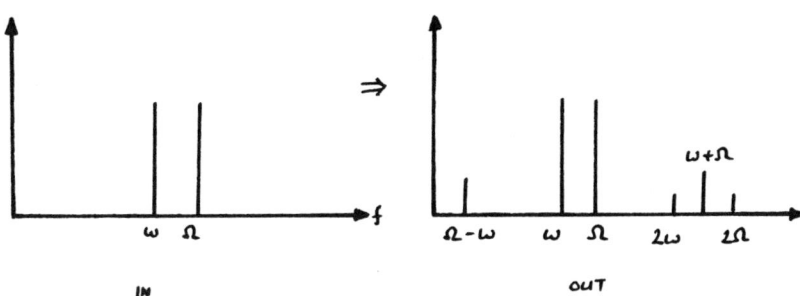

Figure 6.13 Mixing signals to produce an intermediate frequency signal

6.2.7 The Superhet Receiver

To construct a high quality A.M. T.R.F. receiver requires the accurate tracking of a large number of R.F. filters. This is not likely to occur in a real circuit. All T.R.F. receivers contain large numbers of trimmer capacitors and inductors which have to be separately adjusted to make sure that the filters all pick out the same frequencies, and even this does not remove the problem of tracking errors with a large number of variable capacitors.

The *superhet* erodine receiver is the alternative. In this one tuned stage and a *local oscillator* are used. The L.O. frequency is tracked with the tuned circuit frequency in such a manner that the difference between the tuned circuit frequency and the L.O. frequency is a constant, usually 470 kHz or 10.7 MHz. These signals are *mixed* in an amplifier with 2nd order gain, and the output is filtered to remove all but a narrow band around this *intermediate frequency*. The point of this is that the I.F. possesses the modulation of the original signal, and that the carrier frequency is known to a fair degree of accuracy

241

i.e. 470 kHz. Commercially made filters can be bought which are designed to be used in I.F. tuned amplifiers. They are usually produced inside small metal cans, for shielding, and trimmed through a slot in the top.

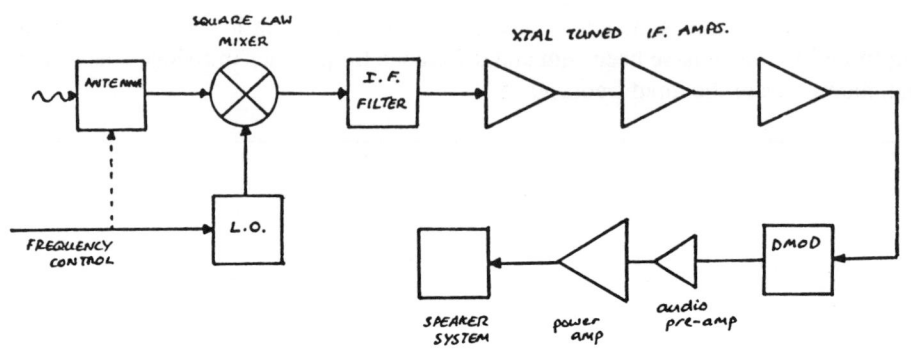

Figure 6.14 A superhet receiver

6.3 Fibre Optic Communication

In this section I shall examine optical fibre communication. Because of the nature of the subject, many of the features of optical fibre communication cannot be described at this level. The optical fibre is a *waveguide*, and a correct treatment of waveguides requires a knowledge of vector calculus as applied, after Maxwell, to electromagnetism.

I shall use *Snell's law*, which is an empirical law from optics, to discuss light propagation in optical fibres, and, to describe the operation of opto-electronic devices, I shall assume that the reader is familiar with the concept of the *photon* as a *particle* of light. I hope that the treatment of light as a wave, or as a particle, whichever is convenient, is not too confusing - it is a fact of life that everything has this dual nature.

6.3.1 Semiconductor LASERs

The LED is a very simple device, and its principle of operation has been explained in §2.2, so I am dedicating this section to a study of the semiconductor LASER.

Light Emission by Excited Atoms

In an atom an electron may exist in any of an infinite number of discrete energy levels.

242

Usually, electrons will end up in those levels with the lowest energy. The highest energy occupied level is called the *ground state*. An electron can be made to jump into a higher energy level by absorbing energy from, and destroying, a passing photon (for which the energy is given by $h\upsilon$; h is *Planck's constant* $\approx 6.63 \times 10^{-34}$ J s^{-1}, and υ is the frequency of the photon) or by collision with other electrons/atoms. The energy of the absorbed photon must exactly equal the difference between the ground state and the higher state that it enters.

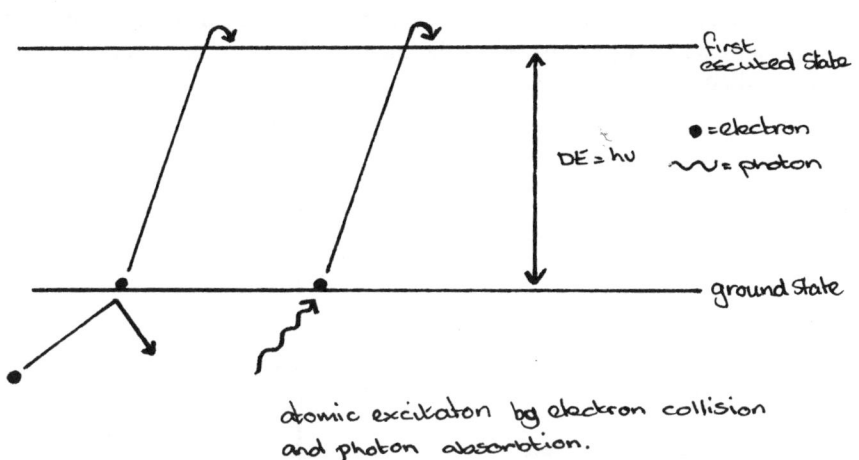

Figure 6.15 Stimulated absorption

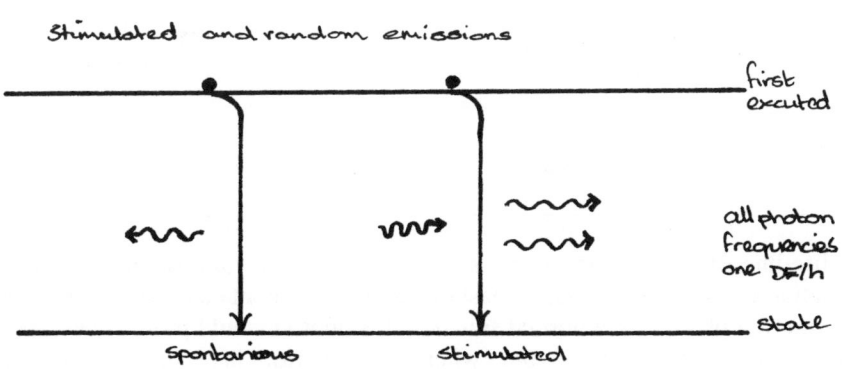

Figure 6.16 Stimulated and random emission

Once the electron is in an excited state it may return to the ground state in one of two ways. It may either *spontaneously decay* into the ground state, or it may be *stimulated* into decaying into the ground state by the passage of a photon, of the same frequency as that absorbed, through the atom. So that it has the correct energy to re-enter the ground state, the electron sheds its excess energy by emitting a photon. A random photon can have any phase, but a stimulated photon will always be exactly in phase with the *stimulating* photon.

Population Inversions and LASER Action

The lifetime of an excited state is entirely random, but the mean value of the lifetime can be predicted. If atoms can be forced into a *long lifetime* state they are likely to stay there much longer than is usual, and so the majority of atoms in a particular sample can be put into an excited state, instead of the majority being in the ground state. Such a situation is called a *population inversion*.

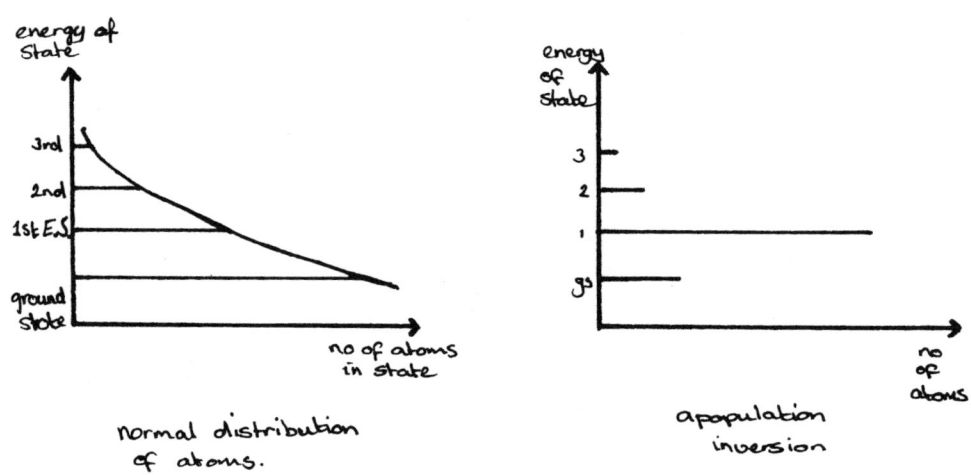

Figure 6.17 A population inversion

The lifetime of any excited state is very short and so, almost immediately after a population inversion is set up, at least one atom will spontaneously decay, causing the emission of a photon. In a normal situation this photon would probably escape without causing stimulated emission but, when a population inversion exists, stimulated emission is almost certain to occur. The initial photon rapidly becomes 2 4 8 ... photons, and a pulse of monochromatic light is emitted by the atoms. This process is called *light amplification by stimulated emission of radiation*.

To make a LASER *cavity* the set of atoms which will 'lase' is placed in an enclosure with a half silvered mirror at one end and a totally silvered mirror at the other. The rest of the sides are made opaque. The mirrors are used to redirect the *avalanche* of photons back into the cavity, where they cause more emission. LASER radiation emerges as a very narrow beam from the half-silvered mirror.

A *pulsed* LASER is one in which the population inversion is created, and then decays immediately, giving a very high power burst of radiation. More useful in communication applications is a LASER in which the population inversion can be continuously replenished and so a lower power, but continuous, beam of radiation is emitted. Such a situation can be created using a forward biased pn-junction.

The Semiconductor LASER

A pn-junction will lase in the correct situations, but it is quite common to form a ppn-junction if LASER action is required. The semiconductor LASER is continuously *pumped* by the forward current of the diode formed, and so emits a beam of infra-red radiation. This beam can be turned on and off by taking the current above or below the *threshold current* of the device. This is the current above which lasing occurs.

6.3.2 Optoelectronic Detectors

Photodiodes

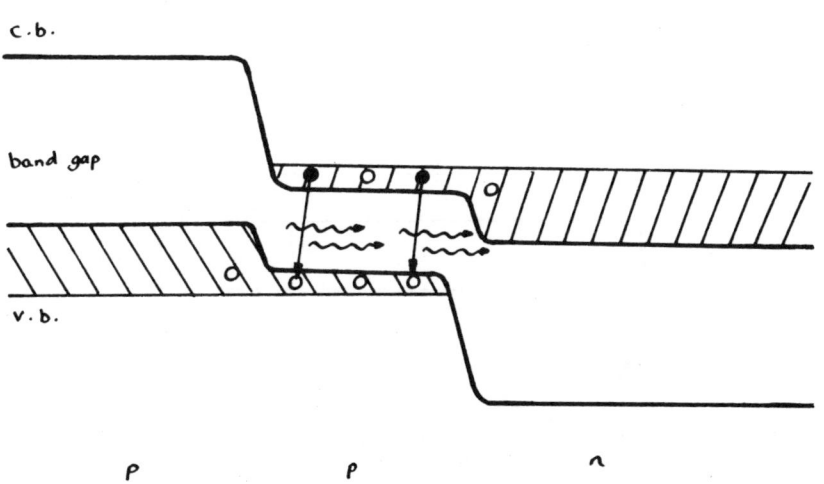

Figure 6.18 Band structure in a PPN laser diode

When a photon, of the correct frequency, is absorbed by an electron in the valance band of a semiconductor the electron jumps into the conduction band and so the effect is the creation of an electron-hole pair. If such pair-production occurs in the depletion layer of a reverse biased pn-junction the large potential barrier causes the electron to be swept into the cathode region and the hole into the anode region of the diode. The reverse leakage current of a diode will therefore drastically increase when the junction is exposed to radiation of the correct frequencies.

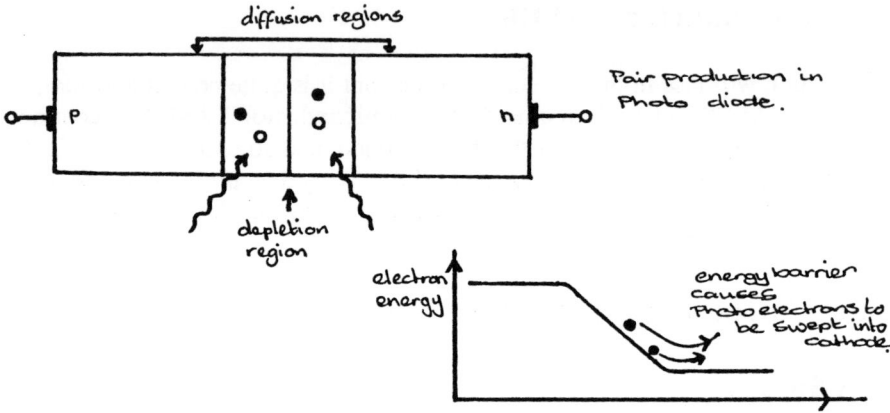

Figure 6.19 The PN photodiode

PIN Photodiodes

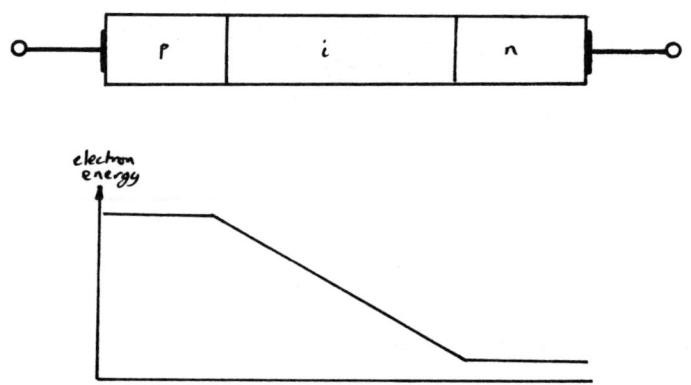

Figure 6.20 The PIN photodiode

Pair production may occur in both the depletion region and the *diffusion region* of the diode, but only photoelectrons/holes created in the depletion region will be rapidly swept to the cathode/anode and cause a significant increase in reverse current. For improved operation of the photodiode, particularly at low wavelengths where photons travel deeper into the crystal before being absorbed, a large depletion region is required. This is produced by reducing the doping to a level where the material behaves as an intrinsic semiconductor and a pin-junction is formed.

Avalanche Photodiodes

Very low light levels are often encountered with long distance optical fibre communications, and a photodetector with internal gain is very useful. The avalanche photodiode, or APD, is a npip+ structure which is held in very high reverse bias (100-400 V) so that photoelectrons acquire a high energy from the field. These high energy *primary* electrons then collide with electrons and cause the creation of *secondary* electrons and holes. The carrier multiplication factor of an APD is often higher than 1000.

Phototransistors

When an incident photon causes pair production in the collector-base depletion region, of a npn-transistor, a hole is generated and swept into the base. The increased base hole density allows more electrons to cross the base-emitter junction, and so the base current increases and the collector current increases dramatically. The phototransistor gives detection with internal gain without the problems of an APD.

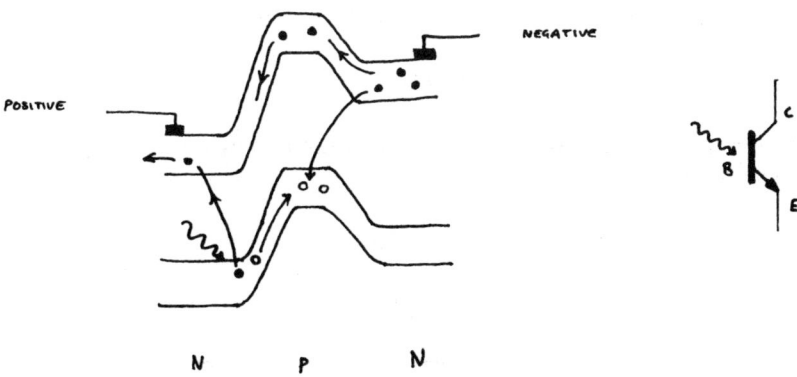

Figure 6.21 Phototransistor Operation and Symbol

Because APDs need high reverse voltages, are expensive to make, are inefficient at low wavelengths, are noisy because of the random nature of the gain mechanism and have a low *quantum efficiency* (defined later) at long wavelengths, there has been interest in the use of phototransistors as photodetectors with internal gain.

Quantum Efficiency and Responsivity of an Optoelectronic Detector

The *quantum efficiency* is defined at the fraction of the incident photons which are absorbed by the detector and cause collection of an electron at the detector's terminals.

$$\eta = \frac{\text{no. of collected electrons}}{\text{no. of incident photons}} \tag{6.2}$$

It makes more sense to treat this quantity in terms of rates, and multiplying both the denominator and numerator of (6.2) by 'per unit time' allows the expression of η in terms of the electron collection rate and incident photon rate.

$$\eta = r_e/r_p \tag{6.3}$$

It should be noted that η is a function of photon energy and so must be quoted at a particular wavelength or frequency.

The *responsivity*, R, can also be useful quantity when discussing the performance of a detector. It is the ratio of output *photocurrent* to incident optical power.

$$R = I_P/P_O \tag{6.4}$$

The incident photon power is the photon rate multiplied by the energy per photon.

i.e.
$$P_O = r_p h\upsilon$$

The photocurrent is the electron collection rate multiplied by the electronic charge, and so I_P is given by (6.5).

$$I_P = \eta \, P_O q/h\upsilon \tag{6.5}$$

$$\therefore \qquad R = \eta \, q/h\upsilon \tag{6.6}$$

6.3.3 Optical Fibre Communication

Optical fibre communication is possible because what is called *total internal reflection* can occur inside a glass fibre. This is a process in which light is reflected along the length of a fibre, without significant leakage of light through the walls of the fibre, and so it is possible to guide light signals through long distances.

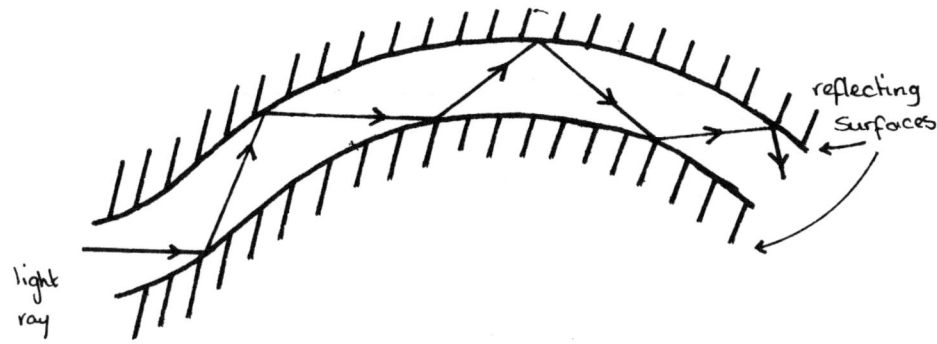

Figure 6.22 Light propagation down an optical fibre

The *refractive index* of a material is defined as the ratio of the speed of light in vacuum to the speed of light in the medium, and is usually represented by the symbol n. It can be shown from electromagnetic and wave theory that when a ray of light is incident upon a *dielectric boundary*, which is a region where the refractive index changes, the ray will split into a reflected ray and a transmitted, or *refracted ray*.

The angles θ_i, θ_t and θ_r obey simple rules known as the *law of reflection* and *Snell's law of refraction*. These rules were discovered experimentally, but can be derived theoretically:

$$\theta_i = \theta_r \quad \text{and} \quad n_1 \sin \theta_i = n_2 \sin \theta_t$$

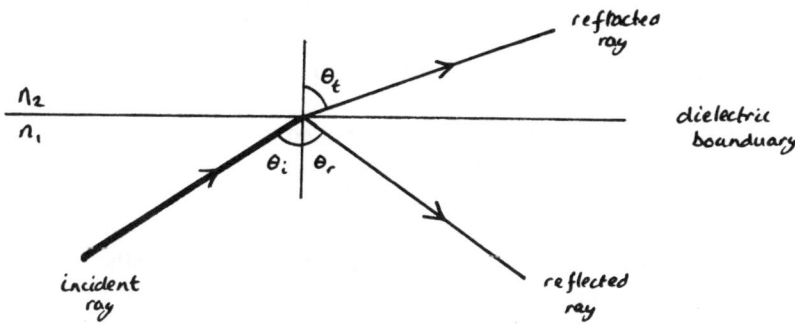

Figure 6.23 Illustrating reflection and refraction

If θ_t exceeds $\frac{1}{2}\pi$ there is no transmitted wave, and the phenomenon of total internal reflection occurs. The condition for this to occur is that the angle of incidence be some critical angle, θ_c, determined from Snell's law as:

$$\theta_c = \sin^{-1} n_2/n_1 \qquad (6.7)$$

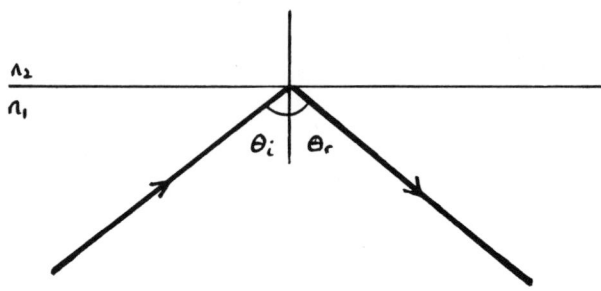

Figure 6.24 Total internal reflection

Methods of Optical Fibre Communication

Optical fibres can be used to transmit light over long distances without significant loss, and have a number of advantages over cable communications.

Some of those advantages are:

Very large bandwidth - near infra-red and visible light occupy a space between approximately 10^{13} and 10^{15} Hz. This gives a signal bandwidth of 9.9×10^{14} Hz, far in excess of that available even to microwave transmissions;

Electrical isolation - the light is transmitted down a fibre which is an insulator, and so does not pick up local electrical noise;

Immunity to crosstalk - it is very easy to ensure that light signals from one fibre do not interfere with signals in another, unlike metallic conductors which radiate R.F. fields every time the signal changes. This also makes optical communication *secure* because the only way to read the data is to fix the end of the cable to a photodetector;

Mechanical advantages - modern fibres have a high tensile strength, and can be bent through large angles without fracture. They are very much lighter than metallic conductors and are much easier to store, transport and install; and

Low transmission losses, with attenuations of less that 1 dB per kilometer, make optical fibre communication very appealing.

The Optical Transmitter and Receiver System

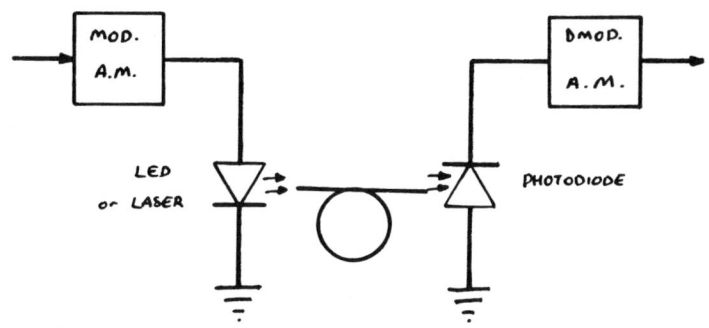

Figure 6.25 An optical fibre communication system

Optical communications are easy to set up, and it is very simple to modulate or demodulate the carrier. A system would consist of an encoder to turn parallel digital data into series data, complete with bit and word synchronisation pulses, and a modulator which merely varies the light output of an LED or LASER according to the data input (i.e. it is an A.M. system). The LED/LASER would then be coupled to an optical fibre. The receiver would be some form of photodetector and a demodulator to restore the original digital data. Alternatively, analogue signals, such as telephone calls, can be sent.

6.3.4 Transmission Properties of Some Optical Fibres

Step Index Fibres

The step index fibre is the simplest type of optical fibre. The refractive index changes abruptly from a high value in the core to a low value in the cladding. The ray reflects from the core boundary along the cable until it reaches the end.

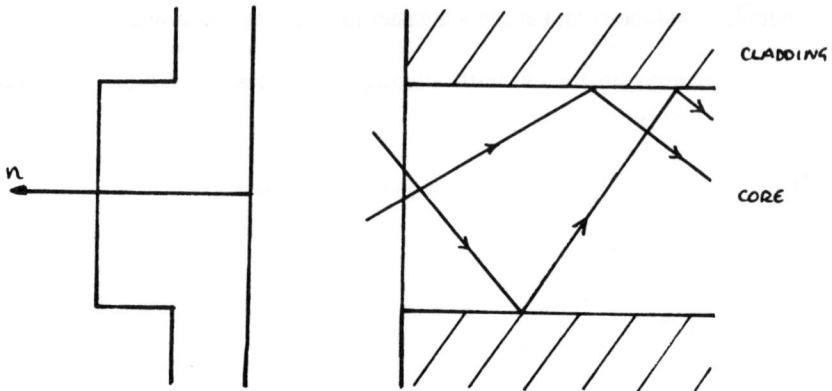

Figure 6.26 Transmission with a multimode step index fibre

Graded Index Fibres

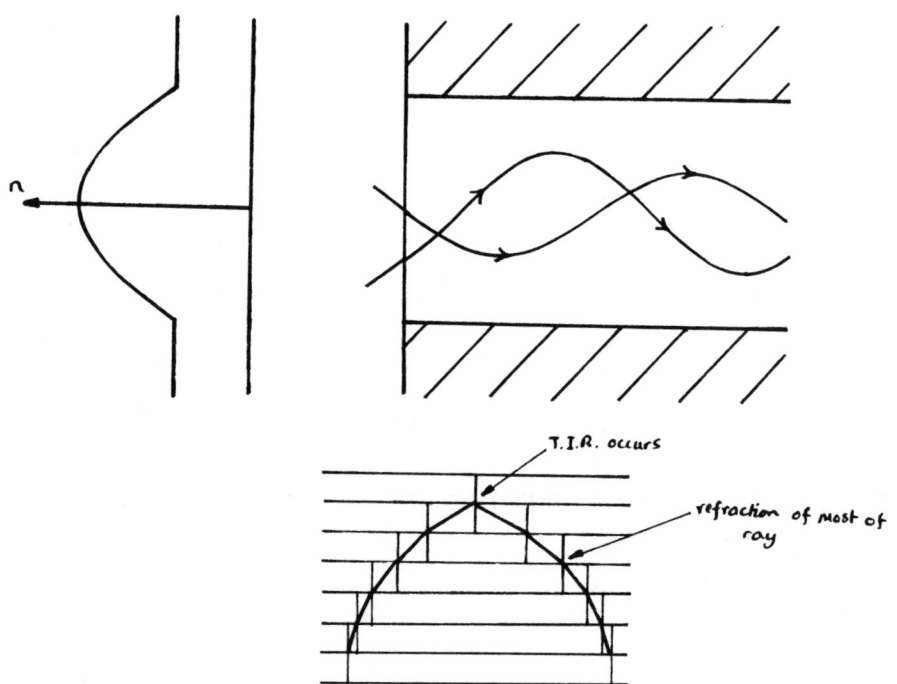

Figure 6.27 Transmission with a graded index fibre

A graded index fibre has a gradual change from a high refractive index in the fibre to a low index in the cladding. This structure means that a ray is gradually bent, as it passes through the fibre, until it totally reflects at some point.

Transmission Loss due to Attenuation

The attenuation of an optical fibre is the ratio of incident power to received (output) power, for a particular fibre, and is usually expressed in decibels per kilometer. A metallic cable has a typical loss of 5 dB km^{-1} whereas modern optical fibres can be made with losses of less than 1 dB km^{-1}, making the optical fibre much more desirable for long distance cable communication.

Absorption

There are also losses due to the atoms of the fibre material absorbing light and entering excited states. The atoms will ultimately re-emit the light, but not necessarily in the direction in which the light was being transmitted, or at the transmitted frequency.

Dispersion

Dispersion is the spreading of a pulse as it travels though a medium.

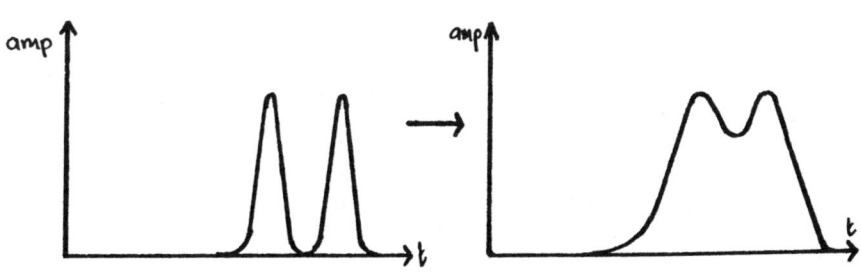

EFFECT of DISPERSION on LIGHT PULSES.

Figure 6.28 The effect of dispersion

The degree of dispersion varies from fibre to fibre, the best results being obtained for *single mode step index* fibres. This fibre contains a core which is only 1 μm wide and which only transmits rays down the length of the fibre. The worst results are for the *multimode step index* fibre, with a wide core, which transmits both direct and reflected rays. Reflected rays take longer to travel down the fibre and, therefore, increase the dispersion to a greater extent than that caused by the properties of the fibre itself. The graded index fibre reduces this problem because the direct ray, in a region of high refractive index, travels with a lower speed than the reflected rays.

Questions for Chapter 6

6.1   Show that the A.M. signal $(1 + v \sin \omega t) \sin \Omega t$ is composed of a carrier frequency, a sum frequency, and a difference frequency. Deduce that the bandwidth required to receive a transmission with an audio bandwidth of 10 kHz.

6.2   A tuned circuit (LCR) is used to couple an aerial to a R.F. receiver. Find the quality factor of the circuit if it is to pass an A.M. audio signal without significant distortion.

6.3   (*harder*) A P.W.M test signal is defined as follows:

$$f(x) = \cos ax, \qquad -\tfrac{1}{2}\pi \leq x \leq \tfrac{1}{2}\pi$$
$$f(x) = 0, \qquad -\pi \leq x \leq -\tfrac{1}{2}\pi$$
$$f(x) = 0, \qquad \tfrac{1}{2}\pi \leq x \leq \pi$$

The function is periodic: $f(x + 2\pi) = f(x)$.

By Fourier series analysis, find the frequency *spectrum* of this signal. What bandwidth is required to receive it without significant distortion?

6.4   A multimode step-index optical fibre has a width μ and refractive index n. A light ray with wavelength λ is travelling down the fibre by total internal reflection. The light is travelling in such a manner that one wavelength fits along the path between reflections. Find the distance, d, travelled by the ray in time, τ. What is the distance, d', that a ray travelling directly down the centre of the fibre would travel in the same time?

6.5   Design a system to transmit and receive 4-bit binary words down an optical fibre. The system must provide both bit and word synchronisation so that the words can be accurately decoded without previous synchronisation of Tx and Rx.

References

Arfken, G, *Mathematical Methods for the Physical Sciences, 3rd. Ed.*, Academic Press Inc.

Bleaney, B.I., and Bleaney, B., *Electricity and Magnetism, 3rd. Ed.*, Oxford University Press.

Blakemore, J. S., *Solid State Physics*, 2nd Ed., Cambridge University Press.

Duffin, W.J., *Electricity and Magnetism*, 3rd. Ed., McGraw-Hill Book Company Ltd.

Eisberg, R.M., *Fundamentals of Modern Physics*, John Wiley and Sons.

Feynman, R.P., Leighton, R.B., and Sands, M, *The Feynman Lectures on Physics*, vol. II, Addison-Wesley Publishing Company.

Horowitz, P., and Hill, W., *The Art of Electronics*, Cambridge University Press.

Millman J., *Microelectronics*, McGraw-Hill Book Company.

Pointon, A.J., *Statistical Physics for Students*, Longman

Riley, K.F., *Mathematical Methods for the Physical Sciences*, Cambridge University Press.

Rosenberg, H.M., *The Solid State*, 3rd. Ed., Oxford University Press.

Stephenson G., *Mathematical Methods for Science Students*, Longman.

Sze, S.M., *Physics of Semiconductor Devices*, 2nd. Ed. John Wiley and Sons.

Sze, S.M., *Semiconductor Devices Physics and Technology*, John Wiley and Sons.

Mathematical Symbols and Abbreviations

$\equiv$	is identical to
$\approx$	is approximately equal to
$\gg$	is much greater than
$\ll$	is much less than
$\propto$	is proportional to
$\rightarrow$	approaches
$\Rightarrow$	is implied by (i.e. 'a $\Rightarrow$ b' means that the truth of statement 'b' is a consequence of the truth of statement 'a')
$\therefore$	therefore
$\|\ \|$	the modulus of, or the magnitude of
$\prod\limits_{i=n}^{N} a_i$	the product of all terms from a_n to a_N
$\sum\limits_{i=n}^{N} a_i$	the sum of all terms from a_n to a_N
$n!$	the product $n(n - 1)(n - 2) \dots 3 \times 2 \times 1$; note: $0! = 1$
$\lim\limits_{x \rightarrow a} f(x)$	the limit of $f(x)$ as x approaches a (i.e. the value to which $f(x)$ tends as x tends to a)
L.H.S.	left hand side
R.H.S.	right hand side
w.r.t.	with respect to

Abbreviations in Use in Electronics

a	anode
a.c.	alternating current
A/D ADC	analogue to digital converter/conversion
a.f.	audio frequency
AGC	automatic gain control

Al		aluminium
ALU		aritmetic logic unit
A.M.		amplitude modulation
As		arsenic/arsenide
BCD		binary coded decimal
b.f.o.		beat frequency oscillator
bit		binary digit
byte		an eight bit binary number
c.c.d.		charge coupled device
c.d.a.		current differencing amplifier
CML		current mode logic = ECL
CMOS		complementary metal-oxide-semiconductor field effect transistor (logic)
CPU		central processing unit
c.r.o.		cathode-ray oscilloscope
c.r.t.		cathode-ray tube
CW		carrier wave keying
D/A	DAC	digital to analogue converter
dB		decibel
d.c.		direct current
DIL		dual in line (package)
DMUX		de-multiplexer or address decoder
d.p.m.		digital panel meter
DTL		diode transistor logic
d.v.m.		digital voltmeter
ECL		emitter coupled logic
EEPROM		electronically eraseable programmable read only memory
EHT		extra high tension (high voltage)
e.m.f.		electromotive force (voltage)
EPROM		eraseable programmable read only memory
FET		field effect transistor

F.M.	frequency modulation
FSD	full scale deflection
FSK	frequency shift keying
Ga	gallium
Ge	germanium
GND	ground (0 V) connection
HEX	hexadecimal (base 16)
h.f.	high frequency
IC	integrated circuit
i.f.	intermediate frequency
IGFET	insulated gate field effect transistor
I/O	input/output
i/p	input
JFET JUGFET	junction gate field effect transistor
k	cathode
LASER	light amplification by stimulated emission of radiation
LDR	light dependent resistor
LED	light emitting diode
LSI	large scale integration
LSTTL	low power Schottky TTL
MASER	microwave amplification by stimulated emission of radiation
MSI	medium scale integration
MOS	metal-oxide-semiconductor
MOSFET	metal-oxide-semiconductor field effect transistor
MUX	multiplexer
μP	microprocessor
n.f.b.	negative feedback
NMOS	n-channel MOS
nybble	a four bit binary number
O	oxygen
o/p	output

op-amp	operational amplifier
osc.	oscillator
o.t.a.	operational transconductance amplifier
P	phosphorous/phosphide
p.c.b.	printed circuit board
p.d.	potential difference (voltage)
PLA	programmable logic array
PLL	phase locked loop
PMOS	p-channel MOS
PROM	programmable read only memory
PSU	power supply unit
p.w.m.	pulse width modulation
RAM	random access memory
r.f.	radio frequency
ROM	read only memory
Rx	receiver
SCR	silicon controlled rectifier (reverse blocking thyristor triode)
S.E.M.	scanning electronic microscope
Si	silicon
SSI	small scale integration
TTL	transistor-transistor logic
Tx	transmitter
UART	universal asynchronous receiver transmitter
u.h.f.	ultra high frequency
UJT	unijunction transistor
USART	universal synchronous/asynchronous receiver transmitter
v.h.f.	very high frequency
v.l.f.	very low frequency
v.c.o.	voltage controlled oscillator
VDU	visual display unit
VMOS	vertical MOSFET

The Système International (S.I.) System of Units

The Base Units

S.I. is a system of units which has been accepted by most nations. The system is based around a set of seven *base units*: the metre (m), the *kilogramme* (kg), the *second* (s), the *ampere* (A), the *kelvin* (K), the *candela* (cd), and the *mole* (mol).

The precise definitions of these units are concerned with basic physics, and need not concern us here. For example, the ampere is defined to be the constant current that, if maintained in two infinitely long wires of negligible cross-section placed one metre apart in vacuum, would produce a force between the wires of 2×10^{-7} newtons per metre length. One newton is the force which causes a mass of one kilogramme to accelerate at a constant rate of one metre per second per second.

The base units used in electronics are the ampere, and the standard "m.k.s." units - the metre, the kilogramme, and the second.

The Supplementary Units

Two supplementary units are used to help in the description of angles.

The *radian* (rad) is the plane angle which is subtended about the centre of a circle by an arc of length equal to the radius of the circle.

The *steradian* (sr) is the solid angle which is subtended about the centre of a sphere by an area, on the surface of the sphere, of magnitude equal to the square of the radius of the sphere. The concept of solid angle may seem a little strange at first, but it is useful in the description of solid figures - as the plane angle is in the description of plane figures. The solid angle subtended about a point by a surface which completely surrounds that point is 4π.

The Prefixes and Multiplication Factors

Physics often requires the study of very large, and very small, quantities. A set of standard prefixes is used to represent various multiplication factors. They are given in table i.

The use of the prefixes *centi* (c = 0.01), *deci* (d = 0.1) and *deca* (da = 10) is discouraged. The use of a comma between groups of three digits is also discouraged (you should use a space as above) because of possible confusion in Europe, where a comma represents a decimal point.

Factor		=		Prefix	symbol
1 000 000 000 000		=	10^{12}	tera	T
1 000 000 000		=	10^{9}	giga	G
1 000 000		=	10^{6}	mega	M
1 000		=	10^{3}	kilo	k
0. 001		=	10^{-3}	milli	m
0. 000 001		=	10^{-6}	micro	μ
0. 000 000 001		=	10^{-9}	nano	n
0. 000 000 000 001		=	10^{-12}	pico	p
0. 000 000 000 000 001		=	10^{-15}	femto	f
0. 000 000 000 000 000 001		=	10^{-18}	atto	a

Table i.

The Derived Units used in Electronics

The *coulomb* (C) is the unit of charge. This is the charge transported per second by a current of one ampere. Defining equation: $I = dQ/dt$

The *volt* (V) is the unit of potential difference. This is the difference in electrical potential between two points of a wire carrying a current of one ampere when the power dissapated by the wire, between the points, is one watt. Defining equation: $P = IV$.

The *ohm* (Ω) is the unit of electrical resistance. This is the resistance between two points of a conductor when a p.d. of one volt applied between the points causes a current of one ampere to flow. Defining equation: $V = IR$.

The *siemens* (S) is the unit of electrical conductance. This is the conductance between two points of a conductor when a current of one ampere flowing through the conductor causes a potential drop of one volt between the points. Defining equation: $I = VG$.

The *henry* (H) is the unit of inductance. This is the inductance of a circuit in which an e.m.f. of one volt is produced when the current flowing through the circuit decreases at a rate of one ampere per second. Defining equation: $E = - L \, dI/dt$.

The *farad* (F) is the unit of capacitance. This is capacitance of a capacitor which has a p.d. of one volt between the plates when a charge of one coulomb has been stored upon the plates. Defining equation: $Q = CV$.

Some Fundamental Constants in S.I. Units

symbol	quantity	value
c	speed of light in vacuo	$2.997\ 294\ 591 \times 10^8$ m s^{-1}
h	Planck's constant	$6.626\ 176 \times 10^{-34}$ J s
e (q)	elementary charge	$1.602\ 198\ 2 \times 10^{-19}$ C
k	Boltzmann's constant	$1.380\ 662 \times 10^{-23}$ J K^{-1}

Table ii

The Decibel

The Bel is a unit used to compare the relative magnitude of two quantities. Those quantites are usually powers. One Bel is a tenfold increase in power. For example, if P_1 = 1 W and P_2 = 10 W then P_2 is one Bel above P_1.

The number of decibels (dB) that P_A is above P_B can be found from the following formula:

$$\text{no. dB} = 10 \log P_A/P_B$$

Index